Forged in War

Forged in War

How a Century of War Created Today's Information Society

R. DAVID LANKES

ROWMAN & LITTLEFIELD
Lanham • Boulder • New York • London

Published by Rowman & Littlefield
An imprint of The Rowman & Littlefield Publishing Group, Inc.
4501 Forbes Boulevard, Suite 200, Lanham, Maryland 20706
www.rowman.com

6 Tinworth Street, London, SE11 5AL, United Kingdom

British Library Cataloguing in Publication Information Available

Library of Congress Cataloging-in-Publication Data

Names: Lankes, R. David, 1970– author.
Title: Forged in war : how a century of war created today's information society / R. David Lankes.
Description: Lanham : Rowman & Littlefield, [2021] | Includes bibliographical references and index. | Summary: "The tools of our information age-from search engines to data mining to smart appliances-grew directly out of conflicts from World War I to the present day. Explore how today's Information Society reflects a worldview shaped by a century of war"— Provided by publisher.
Identifiers: LCCN 2020046487 (print) | LCCN 2020046488 (ebook) | ISBN 9781538148952 (cloth ; permanent paper) | ISBN 9781538148969 (ebook)
Subjects: LCSH: Information technology—History. | Information society.
Classification: LCC T58.5 .L365 2021 (print) | LCC T58.5 (ebook) | DDC 303.48/33—dc23
LC record available at https://lccn.loc.gov/2020046487
LC ebook record available at https://lccn.loc.gov/2020046488

To all those who lost their lives in the global COVID-19 pandemic,
including Luigi Giorno, a beloved husband, father, grandfather, brother, and uncle.
The knowledge infrastructure failed too many, and we must do better.

Contents

Acknowledgments

No one publishes a book alone. Aside from the invisible work that goes into formatting, printing, and distributing, there are those who reviewed drafts and provided feedback. A very special thank you to Wendy Newman, Kim Silk, Bob Stephens, Bet Stephens, Clayton Copeland, Nicole Cooke, and Gina Masullo, whose feedback gave me the needed input to continue on with the project. Thank you to Charles Harmon, whose encouragement and guidance was invaluable. Thank you to the staff of the University of South Carolina Library for the resources, space, and research support. Thanks to Grayce Jones, an amazing graduate assistant and now a fantastic librarian. As always, a special thank you to my patient wife for the early morning writing, and sudden dashes to scribble down ideas.

Introduction

From Alert to a Promise to the Reader

On August 5, 1914, the British ship CS *Alert* left the port of Dover at 1:52 a.m.—just two hours after a British ultimatum to Germany expired, entering the British Empire into the "War to End All Wars." The mission of the *Alert* was the first offensive strike between the two nations.[1] The target of the mission was not a naval base or a warship, but a series of undersea cables. The *Alert* was not a battleship, but rather a cable ship built in 1890 to lay telegraph lines on the seabeds of the world—and, it turns out, to cut German cables under the oceans as well (figure I.1).

As the weather deteriorated, the crew of the *Alert* struggled to cut five undersea cables—cables that connected Germany to France, Spain, and the Azores, and from there to the rest of the world. As the crew contended with weather and failing equipment, a group of approaching destroyers was spotted. The *Alert* was alone. No warships of the Royal Navy could be spared to escort her on her mission. The approaching ships were definitely not British.

The *Alert* refused to retreat and pressed on to cut the fifth and last cable, knowing that in a time of war and without defenses, the decision to stay and complete their mission could mean a quick death. Grapple after grapple went to the seabed to snag the last submarine cable, and returned unusable. Finally, the crew cut the last cable just as the destroyers arrived. Instead of the German Navy, the destroyers were French, and upon learning of the *Alert*'s mission, the French crew cheered on their British allies.

FIGURE I.1
The CS *Alert*. *https://en.wikipedia.ord/wiki/Fike:CS Alert (1).png*

The *Alert*, her crew, and her mission will be central and recurring characters in my look at data, media, and society. Events leading up to the *Alert*'s mission and the effects of those events serve as a unique nexus in understanding our present world. This one cableship represents strategies, technologies, ideas, and actions that still ripple through today's technological, political, economic, and media landscape. Following the threads of history through the *Alert* brings us to today's society with its growing worship of data, a fracturing national narrative, a virtual dismissal of the scientific method in public discourse, growing xenophobic nationalism, and an increasingly corporate higher education sector that replaces students with customers and exploration with assessment.

This work is about searching for a different path from the one forged by the wars of the last 104 years: a renewed humanism that seeks balance between information, theory, and belief in our complex society. It seeks a society where values and human needs drive our views of data, media, and society. It is based on a fundamental understanding that what motivates us—the root of our passion, our genius, our success as people—is learning about the world around us to make meaning in our lives and accumulate power to succeed in those lives. This work is also founded on a belief that we as a society must seek a common good where diverse motivations, passions, and genius are recognized and engaged, and a joint narrative forward can be constructed.

It is an optimistic piece. In this work I will examine the interplay between two vital components of society today: data and media. Data represents the information landscape we build and maintain around us. Media represents the mechanisms we use to communicate and share our knowledge—even when sharing is seen as a force of coercion or manipulation. These two factors—data and media—are not chosen arbitrarily. Even before today's convergence of the internet and the airwaves, both shared common aspirations and rhetoric.

Media and data both have been considered means of liberation and oppression. For every paper extolling the open marketplaces of ideas, there is a Great Firewall of China:[2] a highly censored internet. For all the belief that free speech underpins our democracy, there are parallel discussions of propaganda and controlling the populace through programming and news.

Just as the internet and wide access to information was supposed to make us all smarter, mass media through the century has been presented as an instrument of learning. The printing press, the penny press, the telegraph, telephone, newspapers, radio programs, and television (think Sesame Street) were all supposed to lead to a better, more informed, and better educated society. Yet both the internet and mass media have been accused of dumbing down the public—vast wastelands and distractions for the masses. Cat videos instead of TED talks, *Gilligan's Island* instead of *Masterpiece Theatre*.

Data and media were supposed to bring us together in a global community. Yet both have been demonized as instruments of division and narrowcasting. Filter bubbles and Fox News allow an ever-spiraling series of belief reinforcement and tribalism undaunted by facts, or at least grounded in so-called alternative facts selected to match dogma.

And for all the idealism and public narratives around data and media, there is a competing profit motive that will serve up utopia or dystopia, so long as it is paid for.

To be clear, what is needed is much more than a simple call for a return to some mythical Golden Age of Reason in which facts were worshipped and society held a common vision of the future. No such time ever existed. And, as I'll discuss, facts have never been enough for a society, and "common vision" is often more a result of excluding people by race or economic status or religion or location than of truly building consensus. Also, the very idea of a common vision for a country or society has been shaped by wartime considerations. How has the reality of war, propaganda, and militarization shaped national narratives to this day? How have the wartime needs of our society for domestic morale and foreign alliances molded media, and driven the development of data and science?

My perspective and focus are on the United States, but hopefully connecting to global ideas and developments. In exploring these two pillars of our complex world—data and media—and their intersection, I will touch upon politics, economics, computer science, philosophy, and diversity. However, I am an information scientist, and that is my primary lens of examination. The treatment of the other domains are really just invitations for further investigation and conversation. It is an invitation because no one book, no one person, could ever truly connect all of these dots.

This inability to know everything can be paralyzing. But that's the purpose of this writing. In the face of this complexity, much of society has turned away from the very concept of expertise and replaced it with idolization of belief and the value of the gut. Simultaneously, much of science and scholarship have drawn away from a public culture that has a short attention span and can be cruel—even openly hostile—to the scientific endeavor. Scholars from too many fields focus on increasingly narrow and specialized domains rewarded by systems that emphasize the novel:* new breakthroughs over replication; proving over disproving, even though the cornerstone of any good science is falsifiability.

In the vacuum of respected analysis, *dataism* has risen. Corporations and governments alike now look to make decisions by reading the digital trails left by the modern union of ubiquitous networks, massive-scale computing, machine learning, regulatory frameworks, and vast capital generated by a business model that turns people's behavior into monetized products—a business model so successful, the *Economist* magazine in 2017 declared the data industry more lucrative than the oil industry.[3]

What I am presenting is not a solution, or even a grand plan. Rather I will present a sort of opening statement in a conversation of how we seek to know

* There are some *very* notable exceptions, such as climate science, which is increasingly seeking to engage in a public conversation.

the world around us. To give away the ending, this is a call for smarter communities, a system that respects diversity in all of its forms and shares common goals, and a rejection of a growing worship of data as the sole means of plotting the future. I present a framework for understanding how we make sense of the world, grounded in theory and illustrated with everyday examples. To do this I will outline something I call the *knowledge infrastructure* that exists in all societies, but that today is fragmented and poorly maintained.

FROM HISTORY TO HYSTERIA

Shakespeare wrote, "What's past is prologue."[4] This sentiment drives this book and its structure. I believe that finding a balance between a data-driven world and one founded on individual agency requires an understanding of how our current debate around data and media evolved. While time is a series of events happening sequentially or, as the saying goes, time is just one damn thing after another, history is not. History is the understanding of decisions and choices that were made in the past. It is very much influenced by the present and is dominated by a search for narratives—stories that help us make sense of today by examining yesterday.

Given that, I have organized my thoughts around a series of narrative trips through history, particularly on the events leading up to the First World War until today with some minor excursions further back. I'll examine this history through the lenses of data and media, with each narrative trip a chapter identifying key concepts that shape today's society. It is a 100-plus-year history dominated by conflict and mass national mobilizations: World War I and the advent of mass media, World War II and fascist attempts to rewrite history, a Cold War and the invention of psychological warfare, on through the ongoing cyberwar fed by trillion-dollar tech companies that have blurred the lines between surveillance and a business model founded on collecting and repacking personal data. I sum it up with a look at how all of these ideas play out in the opening months of the coronavirus pandemic.

For example, in chapter 2 I will trace the importance of the telegraph in the early twentieth century to the importance of encryption in today's digitally connected world and the need to make the internet a utility, discussed in chapter 16. In chapter 4 I'll show how the vulnerabilities of the infrastructure supporting the telegraph would lead to the internet itself, and the ramifications of living in a world of always connected devices.

Each trip through history is meant to show the origins of major factors in today's society. This isn't just a conceit for writing. The telegraph network of cables, both on land and under the seas, shaped an amazing amount of core concepts related to the internet of today. Not simply in that these cables allowed global communications that impacted society, but in how traffic was shared across competing and cooperating networks, how data compression was essential to the workings of the network, how encryption was the cornerstone of both commerce and privacy, and ultimately how these developments spurred debates around news and the public purpose of communications.

FROM DICTIONARIES TO DOGMA

It may be helpful to spend some time on definitions for our journey. Many of the terms I use have both a common-sense meaning and a more contentious life in scholarly work. Take the term *data*. In common usage the word is thought of as a set of numbers or observations. Data is also used in a broader sense of facts where "wanting the data" means wanting concrete representations of reality. This ambiguity gets even worse with the term *information*. Is it just another term for data? Are books information? Is it the knowledge that resides in the brains of people?

These questions lie at the heart of chapter 6 and the work of Claude Shannon, but it is worth providing some clarifying usage here. I use the term data in two senses throughout this book. The first is very specific. Data are measurements. Today most of those measurements come from some digital device, such as measurements of temperature, but also measurements of location.

Location in our physical world might be GPS (global positioning system) coordinates. Location in a digital world might be the URL of a web page you're looking at. Data, as we will see, are plentiful. Information is data in context. So, data tells us that the temperature outside is 20°F. Information is knowing that it is cold outside because the measurement is in Fahrenheit and folks find below-freezing temperatures uncomfortable. Knowledge, as we will get into, is knowing that now is a good time to sell mittens in South Carolina. Wisdom is knowing that freezing temperatures in South Carolina are a result of climate change, and that global warming can lead to extreme up-and-down temperature fluctuations.

I also use data as an expansive idea that lies at the heart of dataism—a belief that we can understand and control the world through the accumulation, analysis, protection, and dissemination of data. The word data in this sense is a quick tag to refer to the digital aspects of the world we live in today.

Media is also a tag—a stand-in for a larger construct. It is a term that has different meanings in different historical contexts. In my discussion of the World War I and World War II time periods, I am almost always referring to mass media. That is the use of different communication channels such as newspapers, radio, and television to get a message to a large and diverse audience. However, as I'll discuss, the audience might have been diverse, but that was more because the channels weren't able to more narrowly target specific demographics (and ideologies) than an attempt to be inclusive. So, in the United States during World War II, the audience may have included African Americans and Asian Americans, but they were often not the focus of the message, and certainly had little control over the messages or channels themselves.

This definition of media changes radically when looking at the past 20 years with the advent of broad access to the internet and social media platforms like Facebook. Now traditional mass media producers like advertisers, news organizations, political parties, movie studios, and authors can select their audiences. They can narrowcast messages based on location, ethnicity, ideology, gender, and an amazing array of factors once unimaginable.

Our current understanding of media has also changed with the advent of the internet as a cheap form of expression. Where once the costs involved in sharing a message with millions might have been prohibitively expensive (30-second Superbowl commercials in 2019 cost $5 million[5]), today a viral YouTube video costs pennies over the existing costs of a phone and data plan. To be clear, the costs for that phone and plan are real costs that exclude large parts of the global population. We will be looking critically at the attempts by governments and businesses to overcome those fixed costs, though often at the cost of other things such as privacy and autonomy.

How we define *mass media* is in flux, and has profound implications for a nation seeking a common vision. We are now in the unique situation whereby politicians use the newly fractured media landscape to push a mythical vision of a time when we had a common national narrative (Make American Great Again) in order to sustain an increasingly divided ideological landscape. Yes, that was a twisted and convoluted sentence, but the idea it conveys is even more so: pushing a message of previous unity through channels that sow division for political gain.

Perhaps the most important definition to start us on our path is that of the knowledge infrastructure. The knowledge infrastructure consists of the sources of information, people, technologies, and policies joined by diverse economic motivations that are used by all of us to learn—to build our knowledge. It is not only how we educate members of our society, but how we seek to understand the world and our place in it. The knowledge infrastructure is the Facebook friends that build your feed, and the nightly news show that seeks to encapsulate a complex world into 24 minutes. It is the internet and the copyright laws that attempt to define the rewards of creativity and criminalize remixing. It is advertising and technology industries alike seeking to define your possibilities by the data you produce, and not your voice seeking change. It is the flow of conspiracy theories that feed the debate over mask-wearing during a pandemic. It is the local TV station and local library and Russian chat bots that seek to influence public policy in the United States—just as the British Empire sought to do so in 1914.

It is a system, as I will discuss, that has been with us and evolving over millennia, from the scribes of ancient Egypt, to the invention of the printing press, to today, where much of the infrastructure is digital. It is the conduit of propaganda and the dissenting voice alike. And it is a system that in the past 100 years has been transformed primarily by the needs and investments of war. That transformation has consequences for how you and I live our lives today, and it is one that cannot go unexamined.

FROM DATA TO WISDOM

The structure of the book is inductive by design. That is, rather than starting with a framework and explorations of key concepts followed by examples and our trips through history, we will build the case for reform from the ground up. I feel it is hard to argue against the wholesale redefinition of people as data generators unless you know when and why those choices were made. It is easy to say that there is danger in a business model that treats personal information as currency, and there are plenty of brilliant people doing just that. However, it is not that simple.

Understanding the complexity of new internet business models changes when you know that it is actually the continuation of something started in the 1800s when the penny press transformed readers into consumers and replaced the value of subscribers with the value of eyeballs and advertising.

This then set the stage for mass media and its use to manipulate national policy through propaganda. In fact, the twin tools of media manipulation—propaganda and censorship—take on a whole different dimension when seen in their wartime context. Current conversations on end-to-end encryption and the impact of law enforcement being unable to read the contents of criminals' cell phones takes on a different shade when you know that encryption and the lack of end-to-end secure communication channels is a major reason why the United States entered World War I (see the tale of the Zimmermann Telegram in chapter 1). Our current struggles with misinformation and national priorities during the COVID-19 pandemic take on a new light when you realize that the exact same conversations occurred during World War I and the Spanish flu pandemic (though in that war, people got beat up for NOT wearing masks).

Likewise, the real fear of constant surveillance online needs contextualization before you run to fit a tinfoil hat. What advertisers and hackers can know about your online reality is expansive, and frightening. But like the story of the frog in boiling water (which, by the way, is a myth in popular media—the frog jumps out) we did not arrive at this point all at once. Honest decisions made in the public interest (such as making research literature more accessible, discussed in chapter 6) has led to browser fingerprints and so-called super cookies that power the Amazon advertisement that seems to follow you from site to site (see chapter 5).

If you want a preview of the conclusion, then take this as my thesis statement: a century of war and conflict have weaponized the knowledge infrastructure we use to learn and to control our lives. It is essential that we unbundle concepts of propaganda, a surveillance economy, and a tendency to simplify complex situations to right and wrong, us and them, ally and enemy, in order to embrace a more humane world that rejects the flattening of individuals to data points.

FROM INTRODUCTION TO A COMPACT

We shall begin our case for this new understanding of data, media, and society by focusing on data. What is the information landscape we have created? What does it mean when any piece of trivia is no further away than a glance at your phone, or a literal call into the air to Alexa, or Siri, or Google? And what do we do with the fact that the price for that call into the air is not the cost of the device you purchased, but that every such call can be recorded, dissected, and paid for with our privacy as currency?

One last thing before we jump in: You're right, I really don't know what I'm talking about.

I really should have read that one article. I really should have researched that one topic or gathered that one set of data that is clearly missing. You're right that as a middle-aged white guy who grew up in Ohio, earned university degrees in New York, and now lives in South Carolina, I don't understand the world from the perspective of a woman or a person of color or a European or a native of Asia. I have indeed missed the vital economic theory and the political reality and quite possibly I unconsciously buy into a neoliberal or socialist or capitalist or Marxist ideology. I shouldn't use the term humanism because I'm not an atheist, and talking about society is an act of privilege. Except I have something to say, and I think it is useful in furthering our conversation. And so, if my choice is to seek to share what I do know or stay silent, I chose to share.

Know that I will not cloak these conversations and ideas in a passive voice or scholarly neutrality. These ideas and constructs are mine or are my interpretations of the work of others. That means that when I get things wrong—and I will—I need to learn more. This is, quite frankly, terrifying. Not that I have more to learn—there is true joy in that—but terrifying in that I know that, oftentimes, those whom I need to learn from will be more interested in showing me I am wrong than teaching me what is right.

All I ask is that you approach this work, and these ideas, as a starting point. They need your ideas and your voice as well. You and I, dear reader, must make a compact. I promise to share what I know and be corrected to learn more, and you promise to share what you know in the spirit of learning together. And that sharing begins back on the deck of the *Alert* and the fact that the War to End All Wars started with copper strands that bound the world.

NOTES

1. Gordan Corera, "How Britain Pioneered Cable-Cutting in World War One," *BBC News* online, December 15, 2017, https://www.bbc.com/news/world-europe-42367551; Elizabeth Bruton, "From Australia to Zimmermann: A Brief History of Cable Telegraphy during World War One," last modified September 20, 2013, http://blogs.mhs.ox.ac.uk/innovatingincombat/files/2013/03/Innovating-in-Combat

-educational-resources-telegraph-cable-draft-1.pdf; "CS *Alert* (1890)," *Wikipedia*, last modified September 2, 2019, https://en.wikipedia.org/wiki/CS_Alert_(1890).

2. "The Great Firewall of China," *Bloomberg News*, last modified November 5, 2018, https://www.bloomberg.com/quicktake/great-firewall-of-china.

3. "The World's Most Valuable Resource Is No Longer Oil, but Data," *Economist*, May 6, 2017, https://www.economist.com/leaders/2017/05/06/the-worlds-most-valuable-resource-is-no-longer-oil-but-data.

4. *The Tempest*, Act 2, Scene 1.

5. Nadra Nittle, "What Makes a Super Bowl Ad Successful? An Ad Exec Explains," *Vox*, last modified February 3, 2019, https://www.vox.com/the-goods/2019/1/25/18197609/super-bowl-ads-commercials-doritos-sprint-skittles.

Part I

DATA

If we have data, let's look at the data. If all we have are opinions, let's go with mine.

—Jim Barksdale, former Netscape CEO

Torture the data, and it will confess to anything.

—Ronald Coase, Nobel Laureate

1

Common Carriers

From Telegraphs to Internet Kill Switches

Only one person aboard the CS *Alert* knew its mission upon departure: Superintendent Bourdeaux.[1] He had been expecting the order, as the entire Royal Navy had been deploying for months in anticipation of war. The *Alert* herself had been moved south from port to port and was positioned in Dover for just this night.

The actual order came in the form of a coded telegram. Telegrams may seem quaint to us today, but in 1914 they were the height of technology. The world of telegraphy had skyrocketed since Samuel Morse had linked Baltimore to Washington in 1844 with a physical cable capable of sending electrical impulses in the form of long and short bursts—dashes and dots—Morse code. Up until that point the fastest that information could reliably travel would be by rail or boat.* With the telegraph and its network of land and sea cables, what once would have been months to get a message from Great Britain to Australia took only minutes in 1914.

News, factory orders, and troop reports could circle the globe in under an hour—relayed through a series of telegraph offices and operators clicking away on a single electrical switch—dash after dash, dot after dot. Submarine cables† connected the world, and the primary users of this network were

* The Wright brothers had made their first manned, powered, controlled flight 13 years earlier.

† Submarine cables are communication cables laid along the sea floor—these undersea cables are not to be confused with undersea naval vessels.

government, industry, and the media. Newspapers could now cover the world in close to real time, reporting on events that occurred continents away. Governments used telegrams to more efficiently run sprawling empires and manage complex diplomatic missions. The telegraph revolution didn't just enhance business; it created whole new markets for currency exchange and laid the foundation for multinational corporations.

In fact, governments were instrumental in creating the modern multinational corporation, and this was evident in the expansion of the telegraphy network. Great Britain, Germany, the Netherlands, and the United States all heavily subsidized the telegraph industry—Britain to extend its lead in building a global network of cables and offices, and the rest to try to unseat the British Empire's near-monopoly in telegraphy. The British lead wasn't just a matter of the amount of cables laid around the globe; the Empire had near-exclusive control of cable manufacture and operations.

While much of military history focuses on battles and deployment of soldiers, the beginning of World War I was the first chapter in a new kind of warfare: information war. In addition to cutting German cables in the English Channel, the Royal Navy bombarded ground-based telegraph offices. Throughout the Pacific, Asia, and Africa, colonial and dominion forces* attacked German telegraph stations and cut cables while British diplomats pressured neutral countries to stop relaying German communications.

The vulnerabilities of cables were well known and Germany, in preparation for war, had embarked on a massive construction effort to build a radio network to replace exposed cables. High-powered radio antennas and stations were built on ridges and along coastlines in German territories. However, these were of limited use for three reasons: Germany had little access to the sea, a reality that not only drove the events of World War I but, as we will see, the development of large-scale manufacturing and munitions innovation; radio signals were easy to intercept; and the state of radio technology was less developed than telegraphy. In fact, a primary purpose of the *Alert*'s mission was to force Germany to broadcast its command and control information from the high-powered radio station at Nauen, because it was easier for Great Britain and her allies to intercept radio signals while limiting Germany's effective range.

* These were armed forces made up of local peoples.

The British did not ignore the radio stations,* either. As with the submarine cables, British and dominion forces sought to destroy German radio installations. Radios in German ships docked in neutral countries were locked down or sabotaged. By the end of the first year of the war Germany was virtually cut off from the rest of the world, forced to rely on coded cables easily intercepted by her enemies.

Arguably the most notable victory for this strategy was the Zimmermann Telegram.[2] This was a cable sent along supposedly secure diplomatic lines through the United States to Mexico. The telegram instructed the German ambassador in Mexico to recruit the Mexicans into the war against the United States if the United States joined the war against Germany. Germany would provide funds, and when they won the war, Mexico could reclaim Texas, Arizona, and New Mexico. The British intercepted the message and made it public. It enraged the American population and became a major factor in securing US participation in the war on the side of the British.

It is worth noting that the Mexican government rejected the overtures, in part because it felt Mexico had little chance to challenge the US army, and even less chance of repatriating the non-Spanish-speaking states.

FROM SUBMARINE CABLES TO KNOWLEDGE INFRASTRUCTURE

The telegraphy network was a key aspect of the world's knowledge infrastructure: the people, technologies, policies, and sources of information that a community, nation and, ultimately, a society uses to learn about the world in order to act.[3] Learning, in this case, goes well beyond schools and colleges to how presidents and bankers alike come to develop their view of the world. This worldview is constructed by knowing about everyday local events and major developments across the globe. It is how governments keep informed on the effects of policy and on the clandestine activities of rival states. It is how artists gauge culture in their work. Whereas the transportation infrastructure is the means for going places, the knowledge infrastructure is the means for *knowing* places.

During World War I telegrams, the offices that sent them, and the cables that carried them were a key part of the knowledge infrastructure. They were

* It should be noted that radio at this point in time could be more accurately called wireless telegraphy. Radio signals were not transmitting voice, but coded messages like the telegraph systems with Morse code.

key not just because of the people sending and receiving messages, but the other sources and technologies that fed off of them. Telegrams fed wire services[4]—pooled reporting in place since the Civil War—that would be picked up by local newspapers. Messages over copper lines helped dictators and democrats alike determine policy and allocate resources such as tax revenue.

The knowledge infrastructure at the beginning of the First World War, as it is today, was a mix of public and private investment. In the case of the telegraph, nations were blurring the line between government and private enterprise with massive subsidies to telegraphy companies and to the emergent wireless systems (radio). This mix then, as now, created tensions.

An example of this tension is evident in the business practices of Guglielmo Marconi. Marconi was one of the first to see the wide potential of radio. While today we think of radio as a source of talk and music, at that time a better name might have been wireless telegraph. Marconi didn't invent radio, though he did massively improve radio range and capabilities, and he didn't just sell radios. Today we would say that Marconi provided wireless telegraphy as a service. He owned not only transmitting towers, but often the radios installed aboard ships—the primary market for radio at the turn of the twentieth century. He hired the shipboard radio operators and refused to relay messages from rival companies, locking shippers and navies alike into his company.

Marconi set up subsidiaries in multiple countries, but his business was seen primarily as an English enterprise. Nations like Germany and others wary of extending the British domination of wired telephony to the wireless world formed the International Radiograph Convention in 1906 to create international law for radio operations. The primary purpose of the convention was to force interoperability—that is, Marconi's operators would be forced to relay messages from all radio operators no matter their origin. Radio would be seen as a "public service."

The convention had little effect, as the United States, Great Britain, and four other countries refused to sign on, not wanting to disrupt Marconi's monopoly. It wasn't until the outbreak of World War I that the laws took true effect when the United States signed on and enforced them. The Americans needed radio technology, and they needed open channels of communications for the war effort. This is where the policy aspect of the knowledge infrastructure plays a role. International treaties and national common carrier laws were passed that opened up radio as a market, and as a technology.[5]

Marconi's approach, and the role of government and commerce in 1914, is a clear example of why history is important when seeking to understand today's complex data and media landscape. Aspects of today's knowledge infrastructure and the debate on how to regulate it clearly echo the past. How much exclusive control of the internet can Google have? When Apple created its App Store for iPhones, taking a percentage of every sale, and dictating terms for app design and capabilities, was it a modern Marconi deciding how hardware could be used by limiting interoperability?

An even closer analogy can be drawn regarding nations and the internet. Russia, China, and Iran all have insisted on internet kill switches—the ability to monitor and disrupt traffic within their national borders. Modern-day cable companies such as Comcast have long sought to own not only the network that shares television programming, but the hardware on each end as well—to dictate your cable box. It took government regulation to force cable companies to allow customers to own their own equipment. And as we'll see in chapter 9, cable companies are very much following Marconi's approach to content—seeking to own exclusive rights to the content on the network as well.

The forced interoperability of radio technologies (and operators) is the first example of how our modern data and media landscape was shaped by war. Concepts of common carriers and national regulation of the airwaves was not put in place by international desire, but wartime necessity. The definition of public service and common carriers for radio was initially a military service.

My path forward in the chapters that follow is to flesh out the concepts and reality of an evolving state of our knowledge infrastructure and describe how that evolution was driven by conflict. I begin by examining how Germany's seeming communication isolation at the start of World War I led to a vital piece of our current infrastructure: encryption.

NOTES

1. Gordon Corera, *Intercept: The Secret History of Computers and Spies* (London: Weidenfeld & Nicolson, 2016).

2. "Zimmermann Telegram," *Wikipedia*, last modified June 8, 2020, https://en.wikipedia.org/w/index.php?title=Zimmermann_Telegram&oldid=936231610.

3. This definition of "knowledge infrastructure" is compatible with, though more expansive than, the academic work of the Center for Knowledge Infrastructures located in the UCLA Department of Information Studies and directed by Christine L. Borgman and funded by the Alfred P. Sloan Foundation. The center's work focuses on the multiple "knowledge infrastructures" used within research and scholarship. This includes research data, academic libraries, scientific publishers, and such. For those interested specifically in higher education and research I highly recommend their work: https://knowledgeinfrastructures.gseis.ucla.edu.

4. This is how wire services got their name, by literally sending the news through copper telegraph wires. Susan R. Brooker-Gross, "News Wire Services in the Nineteenth-Century United States," *Journal of Historical Geography* 7, no. 2 (1981): 167–79, https://doi.org/10.1016/0305-7488(81)90119-5.

5. Linwood S. Howeth, *History of Communications-Electronics in the United States Navy* (Washington, DC: U.S. Government Printing Office, 1963).

2

Encryption

From Zimmermann to the Monetized Self

Once the German cable and radio infrastructure was taken out of commission, the Germans didn't just stop sending messages beyond their borders. They continued to use the global telegraphy system. They sent messages through neutral countries, including the United States, at the outset of World War I. In fact, the Zimmerman Telegram (figures 2.1 and 2.2) was intercepted by the British on a supposedly secure U.S. telegraph network that happened to run through London and Canada.

The telegram, in which Germany sought to enlist Mexico, and the resulting outrage from American citizens points out a truism in modern communications: all messages can be intercepted. The security community works very hard to prevent such interception, but during World War I, when your three primary means of transmitting messages were people (who could be tracked, detained, and interrogated), telegraphs (that could be intercepted at relay points and cable offices), and radio (that could literally be plucked from the air), you had to assume that someone other than the intended receiver could get the message.

So, the Germans and the Brits and the Americans, and—well, everyone—sought to make the intercepted message unreadable to all but the intended receiver. The idea of coded messages is probably as old as man's first use of language. Cryptography—the art and science of writing and breaking codes—has always been a part of warfare. The Germans would code their messages

Send the following telegram, subject to the terms on back hereof, which are hereby agreed to

via Galveston

JAN 19 1917

GERMAN LEGATION

MEXICO CITY

130 13042 13401 8501 115 3528 416 17214 6491 11310
18147 18222 21560 10247 11518 23677 13605 3494 14936
98092 5905 11311 10392 10371 0302 21290 5161 39695
23571 17504 11269 18276 18101 0317 0228 17694 4473
22284 22200 19452 21589 67893 5569 13918 8958 12137
1333 4725 4458 5905 17166 13851 4458 17149 14471 6706
13850 12224 6929 14991 7382 15857 67893 14218 36477
5870 17553 67893 5870 5454 16102 15217 22801 17138
21001 17388 7446 23638 18222 6719 14331 15021 23845
3156 23552 22096 21604 4797 9497 22464 20855 4377
23610 18140 22260 5905 13347 20420 39689 13732 20667
6929 5275 18507 52262 1340 22049 13339 11265 22295
10439 14814 4178 6992 8784 7632 7357 6926 52262 11267
21100 21272 9346 9559 22464 15874 18502 18500 15857
2188 5376 7381 98092 16127 13486 9350 9220 76036 14219
5144 2831 17920 11347 17142 11264 7667 7762 15099 9110
10482 97556 3569 3670

BERNSTORFF.

Charge German Embassy.

FIGURE 2.1

The Zimmermann Telegram, as received by the German Ambassador to Mexico. Note that the telegram is encoded into numbers in an attempt to prevent interception. *"The Zimmerman Telegram," National Archives, last modified June 1, 2020,. https://www.archives.gov/education/lessons/zimmermann*

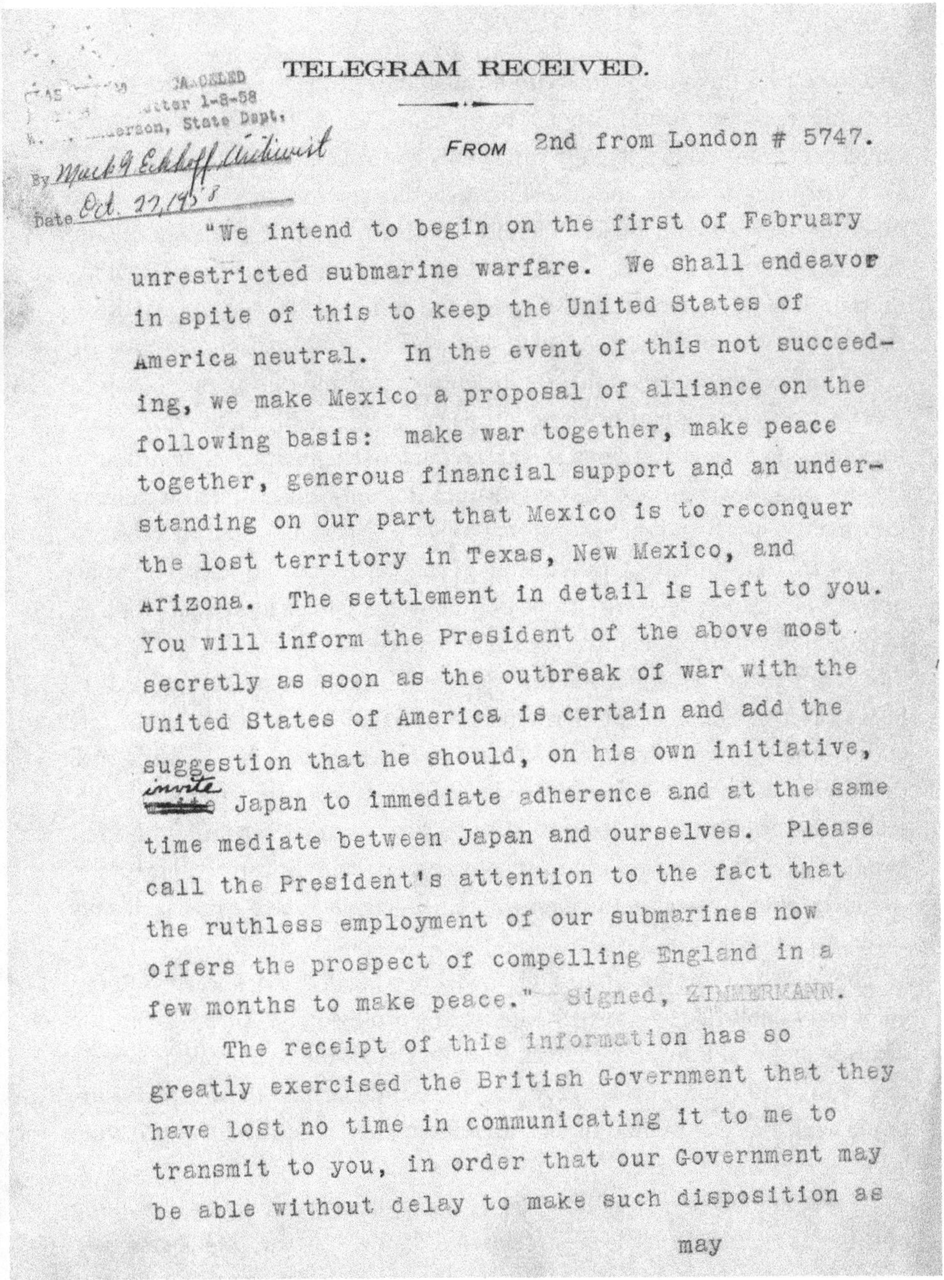

TELEGRAM RECEIVED.

FROM 2nd from London # 5747.

"We intend to begin on the first of February unrestricted submarine warfare. We shall endeavor in spite of this to keep the United States of America neutral. In the event of this not succeeding, we make Mexico a proposal of alliance on the following basis: make war together, make peace together, generous financial support and an understanding on our part that Mexico is to reconquer the lost territory in Texas, New Mexico, and Arizona. The settlement in detail is left to you. You will inform the President of the above most secretly as soon as the outbreak of war with the United States of America is certain and add the suggestion that he should, on his own initiative, invite Japan to immediate adherence and at the same time mediate between Japan and ourselves. Please call the President's attention to the fact that the ruthless employment of our submarines now offers the prospect of compelling England in a few months to make peace." Signed, ZIMMERMANN.

The receipt of this information has so greatly exercised the British Government that they have lost no time in communicating it to me to transmit to you, in order that our Government may be able without delay to make such disposition as

may

FIGURE 2.2
The first page of the Zimmermann Telegram, decoded and translated from German into English. *"The Zimmerman Telegram," National Archives, last modified June 1, 2020, https://www.archives.gov/education/lessons/zimmermann*

and then send them along unsecured cable lines or over the air. The British not only knew this would happen, but planned for it. They installed people called censors at every telegraph office they controlled. Their job was to read each incoming message and determine whether the message was allowed to carry on to its receiver or was actually a coded transmission and needed to be stopped. And when I say read every incoming message, I mean it: "Fifty thousand messages would pass through the hands of 180 censors at UK offices alone every single day. Another 400 worked in 120 stations overseas. In all, 80 million messages would be subject to censorship during the war."[1]

The censors were part of larger propaganda efforts in the warring countries (and many that were not at war) to control the messages being sent to citizens and others abroad. Censors would not only ensure that no military information was being shared, but that only messages that supported the war effort were transmitted. I'll take this up in chapter 7 when I discuss the evolution of propaganda. For now, let me point out that the censors had a problem: what to do with all of the intercepted German coded messages.

The intent of cutting cables and destroying radio infrastructure was to take away a capability of the German enemy, namely, to easily communicate with troops and operatives in the field and beyond borders. It wasn't until the intercepted coded cables and radio transmissions began to pile up (literally) that the British realized they had a new strategic capability on their hands. While they couldn't stop the Germans communicating, they could use intercepted messages to gain an intelligence advantage—that is, if they could break the codes.

The British Admiralty set up a new codebreaking division, known as Room 40. It was established by a former engineering professor, Sir Alfred Ewing, the director of Naval Education. Ewing initially staffed his codebreaking operation with "staff of the naval colleges Osborne and Dartmouth, who were currently available, due both to the school holidays and to naval students having been sent on active duty."[2]

Throughout World War I the development, deployment, and breaking of codes as an art and a science accelerated. Before the war, intelligence and spycraft were considered functions of diplomacy and focused on human actors. During and after the war, intelligence and cryptography became essential capabilities of a successful military. In World War I the British had the upper hand. They not only had the greatest access to coded intelligence;

they had the mathematicians and the support of the military to advance the art. Room 40 would become a huge effort in World War II when it moved to the English countryside and became the Government Code and Cypher School located at Bletchley Park.

In World War II, ciphers and coded messages became strategic assets. The Germans, now of the Third Reich, were well aware of the losses incurred due to intercepted cables and broken codes in the previous war. They also knew that the British had a formidable corps of codebreakers. They would need to find some advantage in new conflicts. They turned to their world-leading manufacturing capabilities. To see the importance of that, I must take a quick detour to 1871.

FROM ENIGMA TO THE BOMBE

Germany had become a powerhouse in manufacturing in the late 1800s after the unification of the German states in 1871. German-speaking principalities in central Europe came together as one country. One of the first acts of the newly unified country was to ensure its position and future. The original plan was to do this through the construction of a great navy. Germany's goal was to rival the British Royal Navy.

The British Empire expanded and remained strong through the nineteenth century due to its power to navigate and control the seas. Colonies in Asia, Africa, Australia, the Caribbean, and the Americas provided England with raw materials and wealth. Throughout this time, it was the Royal Navy that ensured continued fealty to the crown, safe passage of goods, and uninterrupted trade. Even by the start of the First World War, as colonies sought greater independence from England, the British Empire remained strong and the Royal Navy was clearly the world superpower on the seas.

Try as it might, Germany soon realized that a strong naval strategy would have limited chance of success, and so it focused on building a strong manufacturing sector to become an economic power to rival other European states. It was this rise in manufacturing capability that Germany would turn to in order to gain an edge in the world of cryptography in World War II.

Germans looked to use automation and mechanization to build unbreakable codes. They sought to build a complex machine that could produce a code so strong no one could decode it, at least not in time to use the information. The machine they built (and the code it produced) was called Enigma.

The Enigma machine used a complex series of rotors and circuits to implement a dynamic substitution cipher.[3] Each letter typed on a physical keyboard would be transposed to another letter; if you typed an "A," Enigma might code it as a "B." When the receiver typed an "B" on their Enigma machine, it would turn it back into an "A." This kind of substitution code is a long known and not very secure cipher. Regularities in human language provide ample clues to break these codes. For example, in English, E is the most used letter. If you were to look for the most used letter in the coded message and replace it with an E, you might begin to see other words (like "the") and from there figure out the other letters. In addition to E, T and H are often used together, as well as S and H, and so on.

Enigma, however, didn't just do a simple replacement. With every letter typed, the rotors and circuits would change the transposition scheme. So, typing "AA" would not result in "BB"; it might be "BQ." Only another Enigma machine having the same starting conditions could reverse the code . . . or someone with a few centuries on their hands.

In a flash, the British advantage in cryptography was gone. The mathematicians and human techniques for decoding were useless. It would take a new approach to break these codes. It would take using machines to fight machines. At least, that is what Alan Turing thought.

Turing was a mathematician and a genius. He ran a team at Bletchley Park that sought to break Enigma. To do so, his team built a deciphering machine, known as the Bombe:

> This complex machine consisted of approximately 100 rotating drums, 10 miles of wire, and about 1 million soldered connections. The Bombe searched through different possible positions of Enigma's internal wheels, looking for a pattern of keyboard-to-lamp board connections that would turn coded letters into plain German.[4]

The Bombe, and some bad practices of German radio operators, broke the code. It has been estimated that the ability to decode Enigma messages shortened World War II by as much as two years.

Some might call the Bombe a computer, but it was not a computer as we currently know them. It was an analog machine. It had to be hand-wired for different codes. Turing, however, would lay the foundations for the computers of today. But before I go down that track, it is worth looking at the impact that Enigma, the Bombe, and advances in automated cryptography have had on our current-day knowledge infrastructure.

FROM CRACKING CODE BOOKS TO DIGITAL RIGHTS MANAGEMENT

Today we live in a connected world. Whereas the telegraph and submarine cables accelerated message transmission from weeks to minutes, we now expect it in milliseconds. The copper that once bound the word has been replaced by fiber optics—glass cables that transmit data as pulses of light. Whereas early radio used long radio wavelengths to bounce messages off the ionosphere, we now use microwaves to punch through the atmosphere to bounce messages off satellites.

Many of these modern-day advances were fixed in concept by the end of World War II. Take the World Wide Web (I'll explore its development in chapter 5). When the web was first developed, it sent information around the internet in clear open text. These transmissions could be easily intercepted and read. No codebreaking required. As we came to rely on web pages and the internet to do critical functions like read sensitive email, electronic banking, shopping, and even voting, this openness was a problem. It would be relatively simple to pull down credit card numbers, for example. So websites increasingly used encryption to transfer information. Instead of internet addresses starting with "http," more often than not these now start with "https"—that "s" is for "secure." Secure means encrypted.

Today, when you search the web with Google, your queries and results are encrypted. When you buy something on Amazon, your payment information is encrypted. Electronic banking is possible because of encryption.

Encryption has moved from the work of governments to the everyday, and is a fundamental part of our knowledge infrastructure on the internet and beyond. Buy something with a credit card? The machine you use encrypts the information to be sent to your bank. If you use a four-digit code, or your fingerprint, or your face to unlock your phone—you are really providing a key to unlock your encrypted data. Even the music being transmitted from your phone to your Bluetooth headphones is encrypted. If you have purchased a garage door opener in the past decade, the opener is sending an encrypted cycling code to open the door. That keyless button you use to open your car doors? Encrypted.

People often don't take into account encryption when thinking about what we can and cannot do with digital technology. We might think about security with our email or passwords, but what about our music, coffee, and tractors? In society's fast transition to digital media and digital technologies in general, encryption has played a central role in securing our information. It has also redefined what is considered "our" information, and who owns information in general.

In the United States there is a legal principle that we all enjoy, but rarely think about. It is called the "first sale doctrine."[5] Simply put, when you buy some piece of media, you garner certain rights and therefore have a great deal of control over how you "sell, display or otherwise dispose" of that item. If you buy a book, you can read it, lend it out, sell it, or burn it.* Mind you, you only get those rights for that particular physical book or item. So, you can loan out *your* copy of George Orwell's *1984*, but you can't loan out *any* copy of it.

The first sale doctrine has driven a lot of how society works. Libraries in the United States, for example, collect and loan out books based on this doctrine. That's not true around the world. For example, to get a library card in several European countries, you have to pay a yearly fee. Part of that fee goes back to the author of the books you borrowed.

In a pre-digital age, most media consumed would fit under this doctrine. If you bought a bunch of CDs, you could sell them at a garage sale. If you bought an album, you could give it to a friend. However, the wide availability of cheap and fast encryption changed all that when we began to move to digital forms.

In a time when the knowledge infrastructure is overwhelmingly digital, a content producer could choose not to sell you a piece of content, like a book. They could instead sell you partial access to that book in the form of a license. If you have ever "purchased" a song from Apple's iTunes Store or Google's Play Store, you didn't actually buy it. You licensed it. You purchased the ability to listen to that song, but only if you met Apple's or Google's conditions. You could, for example, only listen to that song on an approved device, and only five devices could be approved. This model of being able to access content, but only with the approval and restrictions of

* Don't burn books.

the content creator (or more accurately, the content distributor), is known as digital rights management, or DRM.

DRM is the concept that digital content (and beyond, as I'll discuss in a moment) includes not only the content we access (music, movies, books), but also the software that dictates how we access it. Am I allowed to listen to this song? Yes. Can I copy it for a friend? No. Encryption is the heart of DRM. It prevents you from just ignoring the wishes of the distributor.

DRM and the move from selling things to licensing them affects the television you watch, the music you listen to, and even—and I am not making this up—the coffee you put into your fancy coffee machine. This is what Green Mountain* CEO Brian Kelley was talking about in 2012 when he said this:

> We will be transitioning our lineup of Keurig brewers over fiscal 2014 and early 2015. While we're still not willing to discuss specifics about the platform for competitive reasons, we are confident it delivers game-changing performance. To ensure the system delivers on the promise of excellent quality beverages produced simply and consistently every brew every time, we use interactive technology to help us perfectly brew all Keurig brew packs. Because of this the system will not brew unlicensed packs.[6]

Those "unlicensed packs" Kelley referred to are coffee pods. Keurig created a licensing scheme for coffee in which only their pods would carry a chip that sent an encrypted identifying code to the coffee machine to let the machine brew. DRM for your morning Joe. Encryption allows the knowledge infrastructure's technology to extend to beverages. Where you thought you owned that coffee pod, you only own the ability to make coffee on an approved machine. Marconi would have been thrilled.

Encryption has become so ubiquitous that it is changing the relationship between government and citizens. Where once the resources to build and break codes at a significant level could only be done by governments or large corporations, now it can be everywhere. Where once the police needed a court order to look at your mail or tap your phone, now they also need some way of decrypting the data. That court order only gets them access to the

* In 2012, Keurig was a subsidiary of Green Mountain. Green Mountain changed the company name to Keurig in 2014, and Green Mountain became the brand name of the coffee pods the company produced.

encrypted data. It still requires your password (or finger or face) to actually make use of the data collected.

Encryption has become so fundamental to most digital technologies that it is often invisible. Encryption affects things you might not have known, such as farming equipment. John Deere has made it clear that, while you bought one of its tractors, you did not buy the right to change the software in the tractor's onboard systems.[7] In today's tractors, as in today's cars, the software runs the onboard computer that does everything from timing the firing of sparkplugs to locking the doors. John Deere is saying you own the hardware, but the software is a piece of creative developed content, and therefore protected by DRM and copyright law!

Encryption is scrambling our very concepts of ownership. If you are not in the market for a tractor, how about a sporty sedan like the Tesla Model S? Much has been made of how the car is battery-powered and can drive itself. You can summon the car with your smartphone in a parking lot and it will pick you up. It doesn't need its oil changed, has no transmission gears to break, and you can even play video games with the steering wheel while you wait to power up at a charging station. The vehicle also receives periodic updates over its cellular connection that gives the car new features and improved performance. This all sounds great. But then a hurricane threatened Florida.

In preparation for Hurricane Irma, Tesla sent out a software update to all its cars in Florida. The update gave these cars an extra 30 to 40 miles of range. How was this possible with no new hardware? Simple. Tesla sold different levels of the model S, just like most cars do. The trim (leather, cup holders, and such) wasn't the major difference—the mileage was. Every Model S had the exact same battery pack; Tesla, however, maintained control of the vehicle's software and told the car how much of the battery pack the new owner could use.[8] Buy the lower-priced model, and Tesla's software gave you fewer miles of travel. Our cars are now part of our knowledge infrastructure, and not really completely "our cars."

iPhone owners found out about this new ownership model when their older-model phones started slowing down with new operating systems. Apple knew that older processors in their devices would drain the battery faster with new updates. This could lead to phones quickly running out of juice and turning themselves off without warning. To avoid this bad user interaction, Apple decided to throttle older phones, slowing them down to preserve the battery . . . without telling the owners.

All these examples and the slew of new ownership models (licensing over purchasing, selling hardware but not full access to software, selling hardware capabilities controlled by software) are only possible because of encryption. Encryption that became possible because of the Nazi's adoption of mechanization in cryptography and Turing's understanding that such mechanization could only be countered with mechanization. These events set in motion the spread and wholesale adoption of cryptography and encryption in all aspects of life.

This integration of encryption into the infrastructure of our society shapes how we learn and find meaning in the world. It is a major piece of the knowledge infrastructure, born in war, that we must wrestle with as we move forward. As a society we have to make explicit the link between encryption, privacy, and creativity. We must make choices about where market forces should shape our modern knowledge infrastructure, and where the common good should be preserved or placed above commerce—just as in World War I nations had to choose between a commercial and a public approach to radio.

But perhaps most importantly, cryptography has changed the very notion of ourselves and our personal identity. Encryption is a technical means of securing data: Data such as troop counts or position location of ships at sea. Data such as manufacturing forecasts or details of a corporate acquisition. Increasingly though, data is being used to secure and monetize details about people. Our activities, our accomplishments, our preferences, even our very memories are now able to be represented, to a point, as digital data. That means they can be encrypted, protected, and sold.

FROM THE KITCHEN FRIDGE TO CAPTAIN AMERICA HACKING

I will spend a fair amount of time on how such technologies feed into concepts like privacy and on the difference between public and private spaces later. Here, however, let me briefly explore how encryption allows personal data to be monetized—sold on a market in exchange for goods either by ourselves or by others. We trade our personal data and preferences for services like connecting with friends on Facebook or finding things via Google. In the physical world, we trade our private data for discounts at stores (why do you think you get discounts with your frequent shopper card?). Take two appliances that most people think they bought outright, but in fact have become cheaper because we include our data in exchange for money: televisions and refrigerators.

Televisions are cheap these days. They are getting cheaper every year while also gaining significantly more functionality. We have become used to the idea that technological advancements and economies of scale bring prices down. We are led to believe this is why we can pay less and less for more and more technology over time. But there is a limit to this cycle.

In January 2010, BestBuy.com listed a Samsung 1080p High Definition 46-inch TV for $1,299.99.[9] Quite a deal as this was, according to the site, $1,000 off the regular price (the company must have been clearing extra inventory from the holiday season). What could you get for $1,300 in January 2020 from BestBuy? How about a Samsung 65-inch TV with 4K resolution and built in "smart" features like Wi-Fi and integrated apps to stream shows from Netflix![10] A decade is a long time, but the key here is not the higher resolution or the bigger screen . . . it's the smart features.

The reason you are getting so much more TV for the same money is that you are paying for only a part of the cost of that machine. The rest? Some money comes from content providers like Netflix paying to have a place on your TV and the ability to put ads in front of you. However, there is another significant source of income for the TV manufacturer: you.

There is an app that acts as a source of profits for TV manufacturers, and you probably would never know it was there. In fact, interface designers and manufacturers make it very hard to even find. These apps are called ACRs (Automatic Content Recognition).[11]

Samsung, Vizio, and Roku have built their own ACRs. Sony works with a company called Samba. What do these interface-less apps do? They watch you watch TV. The software uses what flashes over the screen, or what sounds come out of the speakers, to create a digital fingerprint of the content you are viewing. It then transmits these fingerprints every few seconds over that great Wi-Fi feature you thought you bought for your purposes. That information is matched to information on where you live and bundled into a nice package to be sold to advertisers and data aggregators.* Companies sell their TVs at a

* Ironically, given the context of this chapter, not all of this data will be encrypted, so more than just app developers may know what you've been watching.

loss because you are setting them up with an income stream for life: reselling your household's data. Is it any wonder that Samsung is now embedding TV screens and Wi-Fi into their refrigerators?

To be clear, you may well think it is worth sharing your viewing, buying, and eating habits in exchange for watching TV in the kitchen and having artificial intelligence (AI) enabled cameras notify you when you are low on milk. But you need to know that is happening. You are selling a license to your personal data in exchange for a service or item. You are now a content producer. You are now a key player in the knowledge infrastructure as a provider, not just a user or consumer.

What does all of this have to do with encryption? None of this would be possible if acquiring your preferences, your viewing habits, your search history, your grocery buying, and the like was freely available or easy to gather. Marketplaces are built on scarcity, even if artificially created. You demand privacy in, say, your email. A provider ensures you have privacy through encryption. The provider wants to make some more money, so it offers you more features for a reduced price (like free) if it can gather and sell some of your data. If you agree, the company then needs to keep the data it collects secure and scarce so that it can obtain value from it. So, the company encrypts the data (or at least it should). If privacy is so important you may pay the company more, so it won't collect or share data. How does the company provide strong privacy? Encryption.

And make no mistake: encryption is still a vital part of our current ongoing war. I'm not talking about Afghanistan or Iraq, but a cyberwar. It is not just hackers and identity thieves attempting to get your personal data. Governments from China to Russia are actively targeting government and private data sources alike to mine personal data—data that can be used to compromise officials and recruit agents. It may sound a bit like James Bond, but consider the case of a hack of personal data by "Captain America":

> Registering sites in Avengers-themed names is a trademark of a shadowy hacker group believed to have orchestrated some of the most devastating attacks in recent memory. Among them was the infiltration of health insurer Anthem, which resulted in the theft of personal data belonging to nearly 80 million Americans. And though diplomatic sensitivities make US officials reluctant to point fingers, a wealth of evidence ranging from IP addresses to telltale email

> accounts indicates that these hackers are tied to China, whose military allegedly has a 100,000-strong cyberespionage division. (In 2014 a federal grand jury in Pennsylvania indicted five people from one of that division's crews, known as Unit 61398, for stealing trade secrets from companies such as Westinghouse and US Steel; all the defendants remain at large.)[12]

It was the same hacker group that managed to capture millions of records from the U.S. Federal Office of Personnel Management.

Encryption has allowed society to monetize the self. It has fundamentally shifted how the knowledge infrastructure functions, pushing commercial interests well beyond that of government and the public sphere. The adoption of encryption technologies has been enabled by a lack of regulation and a general acceptance (knowingly or not) of monetization of personal data. Further, it has fundamentally shifted the relationship between the consumer and the producer of media. Owners have become subscribers, and content producers have become content landlords, collecting and managing rents.

However, encryption is only a part of this transformation. Someone had to figure out how to make all of this data capturable, how to use the data on a massive scale, and how to build services that were valuable enough to get you to enter into these new transactions. Over the next three chapters I'll describe three fundamental technologies that shape our modern-day data landscape and how each was shaped by U.S. war efforts. I will chart the rise of massive-scale computing, the internet, and how the World Wide Web—extending to phones that now outnumber toilets[13] on the planet—has taken the new business models described in this chapter and supercharged them.

And to do that, I have to return to Alan Turing, but a few years before he decoded Enigma.

NOTES

1. Gordon Corera, "How Britain Pioneered Cable-Cutting in World War One," *BBC News* online, December 15, 2017, https://www.bbc.com/news/world-europe-42367551.

2. "Room 40," *Wikipedia*, last modified May 28, 2020, https://en.wikipedia.org/w/index.php?title=Room_40&oldid=930045890.

3. Karleigh Moore, Ethan W, and Ejun Dean, "Enigma Machine," Brilliant.org, Accessed July 23, 2020, https://brilliant.org/wiki/enigma-machine.

4. B. J. Copeland, "Ultra: Allied Intelligence Project," *Encyclopedia Britannica* online, last modified April 4, 2019, https://www.britannica.com/topic/Ultra-Allied-intelligence-project.

5. "1854. Copyright Infringement—First Sale Doctrine," U.S. Department of Justice Archives, last modified January 17, 2020, https://www.justice.gov/archives/jm/criminal-resource-manual-1854-copyright-infringement-first-sale-doctrine.

6. Graham F. Scott, "The Next Generation of Keurig Single-Serve Brewers Will DRM-Lock Your Coffee," *Canadian Business*, February 12, 2014, https://www.canadianbusiness.com/companies-and-industries/keurig-2-single-serve-coffee-pod-drm.

7. Kyle Wiens and Elizabeth Chamberlain, "John Deere Just Swindled Farmers Out of Their Right to Repair," *Wired*, September 19, 2018, https://www.wired.com/story/john-deere-farmers-right-to-repair.

8. Andrew Liptak, "Tesla Extended the Range of Some Florida Vehicles for Drivers to Escape Hurricane Irma," The Verge, September 10, 2017, https://www.theverge.com/2017/9/10/16283330/tesla-hurricane-irma-update-florida-extend-range-model-s-x-60-60d.

9. "Best Buy Ad: Samsung—46" Class / 1080p / 120Hz / LED-LCD HDTV," Internet Archive, March 4, 2010, https://web.archive.org/web/20100304163725/http://www.bestbuy.com/site/Samsung+-+46%22+Class+/+1080p+/+120Hz+/+LED-LCD+HDTV/9238835.p?id=1218065985439&skuId=9238835.

10. "Samsung—65" Class LED Q70 Series 2160p Smart 4K UHD TV with HDR," Best Buy, accessed July 2, 2020, https://www.bestbuy.com/site/samsung-65-class-led-q70-series-2160p-smart-4k-uhd-tv-with-hdr/6331762.p?skuId=6331762.

11. Read this, please: Geoffrey A. Fowler, "You Watch TV. Your TV Watches Back," *Washington Post*, September 18, 2019, https://www.washingtonpost.com/technology/2019/09/18/you-watch-tv-your-tv-watches-back.

12. Brendan I. Koerner, "Inside the Cyberattack That Shocked the US Government," *Wired*, October 23, 2016, https://www.wired.com/2016/10/inside-cyberattack-shocked-us-government.

13. Tim Worstall, "More People Have Mobile Phones Than Toilets," *Forbes*, March 23, 2013, https://www.forbes.com/sites/timworstall/2013/03/23/more-people-have-mobile-phones-than-toilets/#713d94b76569.

3

Massive-Scale Computing

From Depth Bombs to Deep Learning

Alan Turing was a genius, and his work was instrumental in the development of what we today would recognize as a computer. However, it wasn't his work at Bletchley Park breaking the Enigma code, nor his involvement with the development of the Bombe hardware, that did the codebreaking. Most of the mechanical genius there was supplied by Polish scientists and Harold Keen. Also, the Bombe was most definitely not a computer as we would recognize it today. Instead, Turing's contribution to computing was his foundational thinking about how a universal machine could be developed.

In 1936 Turing published a paper, "On Computable Numbers, with an Application to the Entscheidungsproblem."[1] In it he presented the concept of a "Universal Machine" that could be programmed to solve any mathematical problems and represent varying settings as algorithms. It was this universal machine—now called a universal Turing machine—that would inspire concepts like stored programs, software, and computer memory.[2]

The idea that machines could compute numbers was nothing new. From the abacus to the slide rule to Charles Babbage's mechanical calculator, engineers and mathematicians have always been looking for ways to speed calculations. However, these calculators were limited in that they could only do very general calculations (multiplication, division, addition, subtraction) or were specially made for a given purpose. They were at best tools to speed the work of mathematicians. And mathematicians were busy indeed during

World War II. Aside from the codebreaking work we have already covered, weapons manufacturers needed as many mathematicians as they could get.

If you will recall, Germany after the initial unification in the 1800s dedicated itself to becoming a manufacturing powerhouse. Germany was economically devastated and harshly punished after World War I. It suffered a series of economic collapses including the Great Depression, setting the conditions for the rise of the Nazi Party. The Nazis, once in power, threw off any obligations for reparations or limits to military development forced upon it after the Great War. Military spending quickly re-established Germany's manufacturing sector and prowess.

While the Nazis did not attempt to build a traditional navy of destroyers and battleships to rival the British, they did make major advances in submarines, ground-based weapons, and air warfare. Nazi factories churned out advanced aircraft, tanks, guns, and newer and ever-bigger munitions. This military buildup and expansion only accelerated after another great manufacturing power entered the war—the United States. To be fair, prior to World War II the United States was not considered a manufacturing giant, but prewar investment and mobilization dramatically improved the capability of U.S. factories and engineers. These improvements would help build a middle class and push the United States to become the dominant global economy in the 1950s and 1960s.

What does this have to do with mathematicians? Since the First World War, every ballistic weapon deployed on the battlefield came with something called a firing table. Tank, mortar, cannon, or battleship battery all had a table to tell a soldier how to fire the weapon to hit their target. Why did the soldier need a table? Why not just point the weapon at the enemy and fire? Physics.

FROM CANNONBALLS TO ENIAC

Ballistics is the science of projectiles and firearms. It centers on the fact that no matter how much power put behind a projectile (bullet, cannonball, explosive shell), that projectile will travel in an arc. The force pushing a projectile forward will be countered by gravity pulling the projectile toward the ground. So for any projectile, you need to know not only where a target is, but how far away it is, how heavy your projectile is, and how much force you are putting behind that projectile. Cannonball or lunar lander, the physics are the same. To hit a target at any distance, you have to shoot not at it, but at some angle above it. The

FIGURE 3.1
The gun batteries on a battleship shoot at an upward angle, as seen here on the USS *Iowa*. *https://commons.wikimedia.org/wiki/File:USS_Iowa_(BB-61)_fires_at_North_Korean_target_in_mid-1952.jpg*

higher the angle, up to 45 degrees, the farther the projectile will go. That's why in all the old newsreel footage of destroyers firing at islands or at each other, the guns are facing upward at an angle, not straight out, as shown in figure 3.1.

Here's the trick. In order to know what angle to set your gun to, you must perform literal rocket science. Finding the area under a ballistic arc, for example, requires calculus. There aren't a lot of front-line soldiers that have the ability to do the complex calculations needed, much less under fire. And, as I pointed out earlier, any mechanical calculator was either too limited to perform the necessary calculations, or too specialized to adapt to quickly changing battle conditions.

The solution to this problem was for the arms manufacturers to produce firing tables that would lay out angles for different distances that could be quickly referenced. If you are curious about the work and calculations that go into producing these tables, I refer you to the 109-page report written in 1967 by Elizabeth Dickinson, *The Production of Firing Tables for Cannon Artillery*.[3] I'll be honest, it's not exactly written to keep you on the edge of your seat.

Probable error by the root-mean-square method:

$$PE = S \text{ (Kent-factor) (see Table IV)}$$

where $$S = \sqrt{\frac{\sum_{i=1}^{n} (x_i - \bar{x})^2}{n}}$$

Probable error by successive differences:

$$PE = .4769 \sqrt{\frac{\sum_{i=1}^{n} \delta_i^{\,2}}{n - 1}}$$

where $\delta_i = x_{i+1} - x_i$ $(i = 1, 2, 3 \ldots n-1)$.

FIGURE 3.2
Equations used to compute probable errors in a firing table. *Elizabeth R. Dickinson,* The Production of Firing Tables for Cannon Artillery: Ballistic Research Laboratories Report No. 1371 *(Maryland: Aberdeen Proving Ground, 1967), https://apps.dtic.mil/dtic/tr/fulltext/u2/826735.pdf*

Here's a sample from page 28: "Probable errors are computed in two ways, by the root-mean-square method and by successive differences. Both probable errors are printed, followed by the preferred (smaller) probable error." Dickinson then provided the very helpful equations seen in figure 3.2.

The firing tables were produced by mathematicians working for the manufacturers or the army. As these were the days when drafts and combat were reserved almost exclusively for men, most of the mathematicians making the tables were women, who were called computers.

The problem with producing firing tables is that by World War II, the Allies and the Axis powers were producing weapons and munitions faster than the available pool of computers could keep up. So, the U.S. War Department began looking for some faster mechanical methods for doing the calculations. In the late 1930s while at the Massachusetts Institute of Technology (MIT), Vannevar Bush had developed a mechanical calculator to speed firing table development. It was called the differential analyzer (figure 3.3).[4]

FIGURE 3.3
A differential analyzer, similar to that of Vannevar Bush. *https://upload.wikimedia.org/wikipedia/commons/4/43/Cambridge_differential_analyser.jpg*

While it sped up table development, the differential analyzer was still limited to pre-set conditions, and certainly couldn't be deployed on the fly.* The main problem with Bush's analyzer and other special-built mechanical systems is that they were slow and not adaptable. What was needed was something much more like Turing's universal machine.

John von Neumann and others took Turing's universal machine and began to figure out how a programmable computer could work. Ideas from a century previous on the use of binary mathematics and control, from people like Babbage, George Boole, Ada Lovelace, and Blaise Pascal, lent themselves to new electronic advances like the vacuum tube.

The vacuum tube, though invented in the nineteenth century, advanced rapidly with radio communications during World War I. Vacuum tubes could act as amplifiers and switches, either allowing electricity to flow, or not.

* If you think the differential analyzer in figure 3.3 looks big, Bush built a second version some years later: a 50- to 100-ton analog machine that included 2,000 vacuum tubes and 150 electric motors.

This idea of binary switching—on or off, electrical charge or not—became the basis for the modern computer that we know today.

John Mauchly and J. Presper Eckert of the University of Pennsylvania would put all of these ideas and developments together to build the first digital computer in 1944: the Electronic Numerical Integrator and Calculator (ENIAC) for the U.S. War Department. It filled a 20-foot by 40-foot room and had 18,000 vacuum tubes.[5] Though developed too late to greatly advance the war effort, it was the perfect time for a growing industrial and business sector in the postwar United States.

The postwar era was a remarkable time for computing. However, computers were still limited by their sheer physicality. Not only were they large—they were fragile. The 18,000 vacuum tubes in ENIAC were marvels of modern technology but could break easily. Oftentimes a computing run would have to stop because a vacuum tube broke. More than once the cause was an insect creating a short circuit. This is where we get the modern use of the word "bug" to mean an error in software or technology.

If computers were going to be used beyond the government and the military, new advances would be needed. The key development that ushered in modern computing (and, really, our modern society) came from Bell Labs. Bell Labs had its origins in Alexander Graham Bell and the development of the telephone. The lab grew into a major U.S. think tank working on everything related to telephones and communications. It drew the smartest minds from academia and beyond. In addition to seeking to advance telecommunications, Bell Labs also did work for the U.S. government and pushed forward fundamental physics and mathematics.

FROM VACUUM TUBES TO FLIP PHONES

Work began in 1946 on a replacement for the vacuum tube; this work would later earn William Shockley, John Bardeen, and Walter Brattain a Nobel Prize in 1956.[6] They developed the first working transistor. Like the vacuum tube, a transistor could both amplify electrical signals and act as switches, either allowing or blocking the flow of electricity in a circuit. The major difference was that the transistor was composed of layers of solid silicon so it couldn't shatter. This made computers much more durable.

It is difficult to overstate the importance of transistors on our society. It was not unique as a semiconductor. After all, the vacuum tube was an amazing success in radios and early televisions (and is still used in specialty audio devices). Likewise, the idea of a circuit was not new. The Bombe at Bletchley Park used rotors and gears, but was electrically driven and could be programmed through hardwired circuits. No, the transistor was durable and solid, but most importantly, it could be made small. Small, like a quarter, and then smaller still, like a grain of rice that worked portable radios in the 1970s. Transistors continued to get even smaller—so small, no human hand or tool could make it.

In 1952, Jay W. Lathrop and James R. Nall at the National Bureau of Standards (what would later become the Army Research Lab) developed a technique to mass produce and simultaneously shrink the transistor using photolithography. By doping layers of silicon together and then masking some parts from light, transistors could be exposed and developed like a photograph.[7] Initially hundreds of transistors could be made on a single wafer of silicon, then thousands, then millions only a few molecules wide. In 2017 IBM announced a process that could produce "30 billion transistors packed into a space the size of a fingernail,"[8]

What started as a replacement for ENIAC's 18,000 vacuum tubes became the integrated computer chip or microprocessor. The templates used to photoetch transistors were laid out as complex maps. These maps formed a complicated electrical path that could respond and reroute signals based on stored instructions. These new chips could also hold the memory of the instructions. Soon transistors were being fashioned into specialized chips to process data (central processing units), and chips to store information (random access memory). In more recent times, chips have been developed to run complex graphics (graphics processing chips), and act as radios to link to cell towers and Wi-Fi signals.

The power and precision of the microchip allowed rapid development of new computer accessories such as hard drives and disk drives. Chips controlled precise magnets that hovered microns above iron-coated circular platters that spun at millions of rotations per second. The magnets could read whether the iron was magnetized in certain ways and magnetize the iron when needed. These zones on a disk stored the 1s and 2s that a binary computer needed to operate. For durability, these magnetic zones could be transferred to permanent optical discs such as the compact disc (CD).

The rapid miniaturization of transistors, and the subsequent increasing number of transistors on a chip, made the chip more complex and its thus increased speed was amazingly predictable. Gordon Moore, who helped form Intel, the world's largest chip designer and manufacturer, noted that the number of transistors on the same-sized chip could either double every two years, or the cost of making a chip with the same number of transistors would halve in that timeframe.[9] It became known as Moore's law. It has been applied to other parts of the technology industry as well. The capabilities of a computer double every two years, or the computer costs half as much.

Hard drives and memory have actually beaten Moore's law, doubling even faster. Soon magnets weren't either precise enough or fast enough to keep up with the chips sending data to be read and written. In addition, chip manufacturing became so cheap that transistors and chips were used to replace physical media (hard drives and disks) with solid state drives. The pace of storage development is astounding: "In 1956, the first hard drive to be sold commercially was invented by IBM. This hard drive, shipped with the RAMAC 305 system, was the size of two refrigerators and weighed about a ton. It held 5 megabytes of data, at a cost of $10,000 per megabyte."[10] Today a 128-gigabyte USB "thumb" drive is so cheap, folks often lose them without thinking.

This incredible development from vacuum tubes to transistors to integrated circuits has spawned massive-scale computing on a level that is hard to fathom. Google, the world's largest search engine, runs 15 data centers housing an estimated 2.5 million servers[11] handling 80,173 queries every second.[12] The fastest computers in the world (though to be precise, these are hundreds of computers working together as a single machine) can perform 200,000 trillion calculations per second.[13] Compare that to the British Colossus, another contender for the title of the first digital computer, that could do 100 Boolean queries per second in 1944.

This computing power is clearly a modern-day marvel, but it is not a cost-free advancement for mankind. Those data centers of Google and others consume an enormous amount of power. According to *Forbes*:

> U.S. data centers use more than 90 billion kilowatt-hours of electricity a year, requiring roughly 34 giant (500-megawatt) coal-powered plants. Global data centers used roughly 416 terawatts (4.16 × 1014 watts) (or about 3% of the total electricity) last year, nearly 40% more than the entire United Kingdom. And this consumption will double every four years.[14]

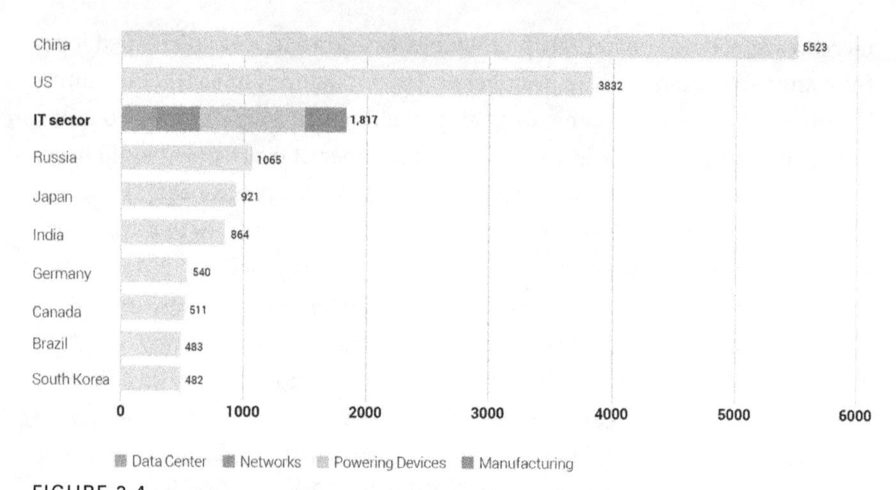

FIGURE 3.4
The power usage of select countries compared to power consumed just by computing. *Used by permission of Anders Andrae*

In fact, many data centers are located near rivers and hydroelectric power stations to assure enough energy to run through those trillions of processes. All that electricity generates heat that must be cooled by air conditioning. That energy signature for computing was estimated by Yale University researchers to be equivalent to the third most power-hungry country in the world—right behind China and the United States, and just ahead of Russia.[15] (figure 3.4).

But what are all of those chips processing? Certainly, we as a species don't ask that many questions. Well, it goes back to the marvels of the transistor. As chips became extremely small, extremely capable, and extremely cheap, businesses, governments and—well, pretty much everyone—put them, well, pretty much everywhere.

Take the obvious. There are now more cell phones in the world than toilets. Each phone, even the flip phones of the 1990s, is full of chips to make calls, save contacts, and play the occasional game of Angry Birds. What you might not know is that if you have an iPhone, or a cell phone that charges using a USB-C cable, there is a chip in the actual wire. The chip controls the electricity to the phone (making sure it doesn't spike) and, of course, enforces which wires can connect to which phones, what data can pass over it using digital rights management (DRM), and the ever-present encryption.

FROM SALTING ROADS TO SMART CITIES

When I talked about your television and your refrigerator (and washer and dryer and car and garage-door opener) using encryption and an internet connection to gather and share data, this was made possible by the power (and size) of microprocessors. You'll find chips in your credit cards and the tags in your car that automatically pay tolls. All of these chips control the most basic function of the knowledge infrastructure: producing data. A lot of these chips busily recording data are also transmitting them as well, so those data centers run by Google, Microsoft, Amazon, Facebook, governments, banks, and so on can keep up-to-date on what the chips are doing and send back instructions if necessary.

The new ubiquity of chips making physical objects "smarter" is seen in the development of the "Internet of Things," or IoT. Once home thermostats were ball-like dials on the wall to manually set a cooling or heating temperature; now you can buy a smart thermostat that can turn down the heat if it detects an empty house, or allow you to turn on the air conditioning via your phone so the house will be cool when you get home. You can buy coffee mugs with an app so you can set the perfect temperature of your beverage. The thermostat and the coffee mug, thanks to the microchip, have joined the knowledge infrastructure.

There is a rush to make just about everything we build "smart," like the roads you ride on. Right now, municipalities all over the world are embedding chips in the roads as sensors with tiny radio transmitters. The goal is to reduce the amount of salt and chemicals put into the environment to thaw frozen roads. Instead of relying on imprecise forecasts, the roads themselves can sense when they are frozen and need to be treated. The data from these sensors are just a minor part of the data our highways are generating.

Soon every mile of road will generate a gigabyte of data per day.[16] This data will come from those sensors embedded into asphalt; real-time traffic data from roadway cameras; and weather information and toll data from radio-frequency identification (RFID) expressway systems, car black boxes, and a myriad of other data sources. It is assumed that this data stream will become a gigabyte an hour as more and more technology finds its way into our vehicles and asphalt (GPS data, real-time environment monitoring, self-driving systems, etc.). As there are 3.5 million miles of highways in the United States, that would be 3.3 petabytes of data per hour, or 28 exabytes per year. To give you a sense of scale, it would only take 5 exabytes to store every word ever spoken by humans.

Roads, thermostats, and refrigerators are all seen to be part of a concept referred to as "Smart Cities"; the idea that with chips in everything from drywall to solar panels, our massive-scale computers could analyze this data and improve efficiency from traffic flow to energy usage. Lights on streets with no people could be turned off. Smart tolls and traffic lights could moderate flow into downtowns to avoid traffic jams. City planners could see with real data where services like police or park rangers should be deployed.

Of course, I write "city planners," but I really mean computers. No human being could possibly make sense of exabytes of data being gathered in real time. We would need a special combination of software and hardware to make this happen, a combination that has been discussed as far back as Charles Babbage's plan for the difference engine in 1819—a combination very much on Alan Turing's mind as he dreamed up the universal machine: a thinking machine. In today's language, we would need Artificial Intelligence (AI).

AI has been the quest of computer science before there was computer science. Turing and contemporaries like Claude Shannon and, later, linguists like Noam Chomsky had been seeking to use all of the growing power of computers to emulate human thinking. A lot of progress was made in the 1990s and 2000s with the advent of natural language processing that used different software techniques to comb through reams of written text, identifying concepts and creating summaries. Much of this work drove early search engines, and indeed present-day services like Google. However, AI always remained an elusive dream. While faster computers could run very complex software, they weren't flexible and were limited in their "human-like" functioning. That all changed when ubiquitous networks (discussed in the next chapter) connected to ubiquitous data-gathering chips and massive-scale computing. Those advances and a new approach to AI known as machine learning are transforming our knowledge infrastructure today.

FROM NATURAL TO ARTIFICIAL INTELLIGENCE

In the 1990s I was working on a project called AskERIC—a service that would answer questions of educators and policy makers online. It was the early days of the web, well before Google, Facebook, or Amazon. Yet even then we would regularly get questions about artificial intelligence (e.g., "Can't machines answer these questions?"). My boss, Mike Eisenberg, had a great answer: "We'll use natural intelligence until artificial intelligence catches up."

A quarter-century later, artificial intelligence has done some significant catching up. From search engines to conversational digital assistance to machine learning embedded in photo apps to identify faces and places, the progress of AI is breathtaking.

Today the larger conversations of AI tend to be either utopian (AI will improve medicine, reduce accidents, and decrease global energy use), or dystopian (AI will destroy jobs, privacy, and freedom). AI has also become a marketing term—soon, I fear, we'll be eating our cereals fortified with AI.

The hype and real progress have merged into a jumbled mess that, overall, can lead to awe and inaction. Awe in that many in the public may feel that the details are over their heads—AI is a game for Google. Inaction because the topic seems too big—what role is there for me when these tools are being created by trillion-dollar corporations?

Here's a true story: the same day my dean asked me about the possibility of creating a degree in data science and AI, MIT announced a billion-dollar plan to create an AI college.[17] I don't think he appreciated me asking him if I would have a billion to work with as well.

I've found these reactions—awe and inaction—are often a result of muddled vocabulary. So, I'd like to briefly break the conversation down into more precise and actionable concepts. Rather than just think of AI as a big amorphous capability, I ask you to think about three interlocking layers: data, algorithms, and machine learning.

The first thing that gets thrown into the AI bucket is the idea of data or big data. As I've described. there is a global uptick in generating and collecting data. With the advent of always-connected digital network devices—smart phones—in the pockets of global citizens, data has become a new type of raw resource.

With this connectivity, most in society have simply accepted (or have been unaware) that one of the costs to being connected is sharing data. Sharing it with a carrier and sharing it with the company who wrote the phone's software. Apple or Google probably know right now where you are, who you're with and, if you use Siri or Google Assistant, they are primed to be listening to what you might be saying right now. No, I mean literally, right now.

The accumulation of data in and of itself is not particularly alarming. As government and corporate bureaucracies have amply demonstrated through time, data means nothing if you don't have systems to find it and use it. This takes us to our second layer of concern in AI: algorithms.

Companies and governments alike are using massive computing power to capture and sort through data, much of it—as I will describe with web bugs and browser fingerprints in chapter 5—identifiable to a single individual. Then these folks make some pretty astounding decisions. Decisions like which ad to show you, or what credit limit to set on your credit card, to what news you see, and even what health care you receive. In our most liberal democracies software is used to influence elections, and who gets interviewed for jobs.

Charles Duhigg, author of *The Power of Habit*,[18] tells the story of an angry father who stormed into a department store to confront the store manager. It seems that the store had been sending his 16-year-old daughter a huge number of coupons for pregnancy-related items: diapers, baby lotion, and such. The father asks the manager if the store is trying to encourage the girl to get pregnant? The manager apologizes to the man and assures him the store will stop immediately. A few days later the manager calls the father to follow up, only to find that the daughter was indeed pregnant, and the store knew it before she told her father.

What's remarkable is that the store knew about the pregnancy without the girl ever telling a soul. The store had determined her condition from looking at what products she was buying, activity on a store credit card, and in crunching through huge amounts of data. If we updated this story from a few years ago, we could add her search history and online shopping habits, even her shopping at other physical stores. It is now common practice to use online tracking, Wi-Fi connection history, and unique identifiers to merge data across a person's entire life and feed them into software algorithms that dictate the information and opportunities they are presented with.

In her book. *Weapons of Math Destruction*,[19] Cathy O'Neil documents story after story of data mining and algorithms that have massive effects on people's lives, even when they show clear negative effects and outright discrimination. She describes investment algorithms that not only missed the coming of the financial crisis of 2008, but actually contributed to it; and models that increased student debt, prison time for minorities, and blackballed those with mental health challenges from jobs. The recurring theme in O'Neil's work is that these systems are normally put in place with the best of intentions, but often have very different consequences.

Here we find the key issue for our understanding of the knowledge infrastructure and the use of software to crunch massive amounts of data to make decisions on commerce, health care, credit, and jail sentences. That issue lies in the assumptions that those who write and use software make—often very dubious, and downright dangerous, assumptions. Assumptions like algorithms are objective, and data collection is somehow a neutral act. Or even the assumption that everything can be represented in a quantitative way, including the benefit a person provides to society.[20]

For too long, too many have seen the act of data collection as neutral. Too many policy makers, professionals, and average citizens assumed that collecting, describing, and providing data were acts either without bias, or with biases controlled. Scholars and data professionals alike now understand that the decisions we make in everything from what we record to whom we record from are choices. Those choices may be guided by best practice, or even enforced by law but, ultimately, they are human choices in a material world where resource decisions must be made.

And this brings us to our last layer—the layer that most purists would say is true artificial intelligence development: the use of software techniques to enable machine learning and the more specific deep learning. That is, software that allows the creation of algorithms and procedures without human intervention. With techniques like neural nets, Bayesian predictors, Markov models, and deep adversarial networks, software sorts through piles and piles of data-seeking patterns and predictive power.

An example of machine learning systems in action would be feeding a system a number of prepared examples, say, hundreds of mammograms[21] that are coded for signs of breast cancer. The software builds models over and over again until they can reproduce the coded results without the prepared examples. The trained system is then set upon vast piles of uncoded data using the new internally developed models to look for breast cancer at a rate, and often at a level of accuracy, that humans can't achieve.

With the wide availability of massive data, newer deep learning techniques do away with the coding, and go straight to iterative learning. Whereas machine learning used hundreds of coded examples, deep learning sets software free on millions and millions of examples with no coded examples—potentially improving the results and eliminating the labor-intensive teaching phase.

When this works well, AI software can be more accurate than humans doing the same tasks—billions of operations per second, finding pixel-by-pixel details humans could never see. And these systems can do it millions and billions of times, never tiring, never getting distracted.

In these AI systems there are two issues that society must respond to in our construction of the knowledge infrastructure. The first is that these machine-generated algorithms are only as good as the data they are fed. Mammograms are one thing; credit risks are quite another. Just as with our human-generated algorithms, these systems are very sensitive to the data they work with.

For example, a maker of bathroom fixtures sold an AI-enhanced soap dispenser. The new dispenser reduced waste because it was extremely accurate at knowing if human hands were put under the dispenser versus, say, a suitcase at an airport. Extremely accurate, so long as the hands belonged to a white person. The system could not recognize darker skin tones.[22] Why? Was the machine racist? Well, not on its own. It turns out it had been trained only on images of Caucasian hands—ultimately a human decision.

We see example after example of machine learning systems that exhibit the worst of our unconscious biases. Chat bots that can be hijacked by racists through Twitter, job-screening software that kicks out non-western names. Image classifiers labeling images of black people as gorillas.[23]

However, bad data ruining a system is nothing new. The real issue here is that the models developed through deep learning are impenetrable. That mammography example looking for breast cancer? The programmers can tell you if the system detected cancer, even the confidence the software has in its prediction. The programmer can't tell you how it arrived at that decision. That's a problem. All of those weapons of math destruction that Cathy O'Neil described can be audited. We can pick apart the results and look for biases and error. In deep learning, everything works until, well, an airplane crashes to the ground or an autonomous car goes off the road.

And so, what are we to do? This is tricky. There can be no doubt that data analytics, algorithms taking advantage of massive data, and AI have provided society with great advantages. Look no further than how Google not only has the ability to search through trillions of web pages in milliseconds, but often serves as a digital document delivery service undreamed of 25 years ago when I was working on AskERIC. Yet, we still need that natural intelligence that Mike Eisenberg talked about.

Our communities and our society need a voice to ensure the data being used is representative of all parts of a community, not just the dominant voice, or the most monetizable. Our communities need support, understanding, and organizing to ensure that the true societal costs of AI are evaluated, not simply the benefits.

That may sound like our job is to be the critic or even the Luddite, holding back progress. But that's not what we need. We all need to become well versed in these technologies, and participate in their development, not simply dismiss them or hamper them. We must not only demonstrate flaws where they exist, but be ready to offer up solutions. Solutions grounded in our values and in the communities to which we belong.

We need to know the difference between facial identification systems, and facial identification systems that are used to track refugees and peaceful protestors. We need to know the difference between systems that filter through terabytes of data, and systems that create filter bubbles that reinforce prejudice and extremism.

If we are to sensibly and deliberately incorporate AI into the knowledge infrastructure, that is, consider it holistically beyond the technical and commercial applications, we need to look beneath the software to the philosophy and methodologies guiding the coders. To understand our individual agency and responsibility, we need to see beyond what AI can do, into what we as individuals, as citizens, as consumers, and as community members want it to do.

In the next chapter I will continue the discussion of foundational knowledge infrastructure technologies by laying out the development of the internet and the World Wide Web. I'll then return to discuss how the work of Claude Shannon formalized the concept of data, leading to another foundational piece of the knowledge infrastructure's technology: data compression, as an introduction into the larger issue of *dataism*—a philosophy that sees the world through the data we can collect and analyze. I will then talk about how the successes of Bell Labs and Shannon were part of a massive growth of what we consider the knowledge infrastructure in the Second World War and beyond, particularly the growth of access to scientific and technical information and increased government funding of research. I will then discuss how this new investment by the federal government has transformed higher education.

It all begins, however, with Little Boy, and what happens when you connect all this computing capability together.

NOTES

1. Alan Turing, "On Computable Numbers, with an Application to the Entscheidungsproblem," *London Mathematical Society* 2, no. 42 (1937): 230–65, http://doi.org/10.1112/plms/s2-42.1.230.

2. Manolis Kamvysselis, "Universal Turing Machine," Web MIT, accessed July 14, 2020, https://web.mit.edu/manoli/turing/www/turing.html.

3. Elizabeth R. Dickinson, *The Production of Firing Tables for Cannon Artillery: Ballistic Research Laboratories Report No. 1371* (Maryland: Aberdeen Proving Ground, 1967), https://apps.dtic.mil/dtic/tr/fulltext/u2/826735.pdf.

4. "Claude Shannon—Information Theory and More," I Programmer, last modified April 27, 2017, https://www.i-programmer.info/history/people/351-claude.html.

5. Kim Ann Zimmermann, "History of Computers: A Brief Timeline," Live Science, September 7, 2017, https://www.livescience.com/20718-computer-history.html.

6. "Transistor," *Wikipedia*, last modified June 20, 2020, https://en.wikipedia.org/wiki/Transistor#:~:text=A%20transistor%20is%20a%20semiconductor,connection%20to%20an%20external%20circuit.

7. For more on the production of microchips using light, see Kevin Bonsor, "How EUVL Chipmaking Works," How Stuff Works, May 24, 2001, https://computer.howstuffworks.com/euvl1.htm.

8. David Nield, "IBM's New Computer Chips Can Fit 30 Billion Transistors on Your Fingertip," Science Alert, June 6, 2017, https://www.sciencealert.com/new-computer-chips-can-fit-30-million-transistors-on-your-fingertip.

9. "Moore's Law," *Wikipedia*, last modified June 30, 2020, https://en.wikipedia.org/w/index.php?title=Moore%27s_law&oldid=937486060.

10. Nigel Harley, "The History of the Hard Drive." Think Computers. June 4, 2013. https://thinkcomputers.org/the-history-of-the-hard-drive.

11. "Google Data Center FAQ," Data Center Knowledge, March 17, 2017, https://www.datacenterknowledge.com/archives/2017/03/16/google-data-center-faq.

12. "Google Search Statistics," Internet Live Stats, accessed July 9, 2020, https://www.internetlivestats.com/google-search-statistics.

13. Micah Singleton, "The World's Fastest Supercomputer Is Back in America," The Verge, June 12, 2018, https://www.theverge.com/circuitbreaker/2018/6/12/17453918/ibm-summit-worlds-fastest-supercomputer-america-department-of-energy.

14. Rado Danilak, "Why Energy Is a Big and Rapidly Growing Problem for Data Centers," *Forbes*, December 15, 2017, https://www.forbes.com/sites/forbestech council/2017/12/15/why-energy-is-a-big-and-rapidly-growing-problem-for-data -centers/#2799dc5a307f.

15. Fred Pearce, "Energy Hogs: Can World's Huge Data Centers Be Made More Efficient?" Yale School of the Environment, April 3, 2018, https://e360.yale.edu/ features/energy-hogs-can-huge-data-centers-be-made-more-efficient.

16. "Transportation Knowledge Networks: A Management Strategy for the 21st Century," Special Report 284, Transportation Research Board (Washington, DC: 2006), www.trb.org/news/blurb_detail.asp?id=5789.

17. Will Knight, "MIT Has Just Announced a $1 Billion Plan to Create a New College for AI," *MIT Technology Review*, October 15, 2018, https://www.technology review.com/f/612293/mit-has-just-announced-a-1-billion-plan-to-create-a-new-col lege-for-ai.

18. Charles Duhigg, *The Power of Habit: Why We Do What We Do in Life and Business* (Toronto: Anchor Canada, 2014).

19. Cathy O'Neil, *Weapons of Math Destruction: How Big Data Increases Inequality and Threatens Democracy* (London: Penguin Books, 2018).

20. Alexandra Ma, "China Has Started Ranking Citizens with a Creepy 'Social Credit' System—Here's What You Can Do Wrong, and the Embarrassing, Demeaning Ways They Can Punish You," Business Insider, October 29, 2018, https://www.businessinsider.com/china-social-credit-system-punishments-and -rewards-explained-2018-4.

21. Karen Hao, "Google's AI Breast Cancer Screening Tool Is Learning to Generalize Across Countries," *MIT Technology Review*, January 3, 2020, https:// www.technologyreview.com/f/615004/googles-ai-breast-cancer-screening-tool-is -learning-to-generalize-across-countries.

22. Tom Hale, "This Viral Video of a Racist Soap Dispenser Reveals a Much, Much Bigger Problem," IFLScience!, August 18, 2017, https://www.iflscience.com/ technology/this-racist-soap-dispenser-reveals-why-diversity-in-tech-is-muchneeded.

23. Jana Kasperkevic, "Google Says Sorry for Racist Auto-Tag in Photo App," *Guardian*, July 1, 2015, https://www.theguardian.com/technology/2015/jul/01/ google-sorry-racist-auto-tag-photo-app.

4

The Internet

From Hydrogen Bombs to 5G

In this chapter and the next, I am going to trace the rise of the internet. However, it is really the story of how early decisions in preserving the command and control capabilities of the US government in an atomic war resulted in digitizing major parts of the knowledge infrastructure, and how early technical decisions in the network's creation have built up entire industries to track us online. This story is crucial for understanding the knowledge infrastructure, because the impact of our computing power, discussed in the previous chapter, is almost completely reliant on networks to connect and scoop up data.

I begin this examination with the *Enola Gay*, a specially modified Boeing B-29 Superfortress, taking off from Tinian in the North Mariana Islands at approximately 2:45 a.m. on August 6, 1945, starting the world's first nuclear war.

The *Enola Gay*'s mission was to fly to Hiroshima, Japan to drop a new type of bomb—an atomic bomb. The bomb was code-named Little Boy.[1] "At 8:15 a.m., the bomb was released over Hiroshima. While some 1,900 feet above the city, Little Boy exploded, killing tens of thousands and causing widespread destruction."[2]

Little Boy was a uranium bomb, and had the destructive power of 15 kilotons—the equivalent destructive power of 15,000 tons of TNT. In the logs of the *Enola Gay* the event is marked with one small comment: "10:52 Cloud Gone."[3]

The only other atomic bomb ever used in war was Fat Man, dropped by the *Bockscar* on Nagasaki, Japan, three days later.[4] Fat Man was a different type of atomic bomb than Little Boy. Instead of uranium it used plutonium, with a yield of 20 kilotons. The horror of the power and destruction of an atomic bomb brought a quick end to the war in the Pacific.

To give you a sense of how much power these weapons and subsequent thermonuclear weapons can unleash, realize that the entire destructive force of *all* the weapons exploded in World War II, including Little Boy and Fat Man as well as the bombing of London and Dresden, is estimated at 3 megatons (a megaton is roughly a 1,000 times more than a kiloton, or 1,000,000 tons of TNT). The current total global nuclear arsenal capacity is estimated around 1,460 megatons or 1.460 gigatons (the explosive energy equal to that of 1 billion tons of TNT).[5]

Unfortunately, the destructive potential of a 20-kiloton bomb wasn't enough to lead to their immediate outlaw. Four years later the Soviet Union exploded a near-replica of the Fat Man bomb at the Semipalatinsk Test Site in Kazakhstan on August 29, 1949, ensuring Russia's place as a world superpower and locking the globe into the Cold War.[6]

The detonation of atomic bombs at the end of World War II would forever change warfare and the knowledge infrastructure, setting off a series of technological developments that impact your life every day. The dropping of the bomb led directly to the development of the internet and the World Wide Web. And the choices made in these developments created a system that simultaneously drained billions of dollars of value in the music industry and led to mass surveillance that makes every site you load on your phone available to advertisers and hackers alike.

FROM LITTLE BOY TO ARPANET

What do atomic bombs have to do with the music industry and the internet? It all started with an effect known as an electromagnetic pulse (EMP) triggered by the nuclear explosions over Japan, and a small military branch in the newly formed U.S. Department of Defense—the Advanced Research Projects Agency, better known as ARPA.* ARPA was a think tank within the Department of Defense, created in response to the Soviets launch of Sputnik, the world's first satellite. In 1960 ARPA was worried about military communications during a possible nuclear war.

---

* In 1972, ARPA was renamed the Defense Advanced Research Projects Agency or DARPA.

When you detonate a nuclear weapon in the atmosphere, it generates a huge amount of energy. Some of that energy is in the form of heat, some in the form of a pressure wave that decimates people and buildings alike. Another type of energy produced is light. Some of that light can be seen in the blinding flash of the detonation. While this visible light can blind an observer, the much more dangerous and deadly light is invisible gamma and x-rays. When gamma rays interact with the air and the ground. they generate an intense broadcast of electrically charged particles that overload and destroy electronic components in radios, or computers, or telephone switches.[7]

ARPA commanders and scientists knew that just as the German communications could be severely compromised by cutting submarine cables, the U.S. phone system could be crippled with one well-placed nuclear blast in the middle of the country. The American telephone system of that time was barely able to provide enough capability for consumer long-distance calls. It was too fragile to handle the communications needs of a major military action on both coasts of the country. The United States needed a more robust way of getting military messages where they needed to go.

Solving the communication problem would help solve another problem the Department of Defense had at the time: paying for computing capacity at universities and research centers conducting military projects. In the 1960s and 1970s computing power was still expensive. Major universities had used government funds to build large computers. ARPA realized that there was no way the agency could afford to build expensive computers (that took up whole floors of buildings) for every university and research center that would need them. ARPA managers wanted to find a way to connect these computers so that a researcher at, say, the University of Chicago could use a computer at Stanford University, 2,000 miles west.

ARPANet was born when the Department of Defense awarded a contract to build a network for connecting research computers to Bolt, Beranek and Newman Inc. (BBN) on April 7, 1969. BBN designed and built small highly specialized computers called Interface Message Processors (IMPs) that were the forerunners of today's routers. ARPANet began as three interconnected military computers in Utah and California. In 1970 the network was extended to BBN's Massachusetts office, and then to several universities. ARPANet continued to grow rapidly. In 1972 there were 24 sites; in 1973, 37 sites, including one in England. 1974: 62 sites. 1977: 111 sites. By 1981, there were 213 host computers on the network with another host connecting approximately every 20 days.[8]

FROM ARPANET TO INTERNET

By 1990 ARPANet had connected most research universities, had gateways to military networks, and extended around the world. It was officially handed off to the U.S. National Science Foundation (NSF) and renamed the NSFNet. However, in 1990 few referred to it as NSFNet, preferring the simpler, and more accurate name: the Internet.

There is no doubt that ARPANet had succeeded in one of its objectives—connecting together computing resources. It had also succeeded spectacularly in its second mission—creating a system that minimized the vulnerabilities of the old telephone system.

Prior to the Internet, the U.S. phone system was a circuit-switched network. When someone placed a call across town, or across the country, a dedicated circuit connected the two phones. This meant that when a lot of people were making calls, you could get a message telling you "all circuits" are busy. If you wanted to add capacity for more calls, you had to run more cables across telephone poles. It also meant that if you cut those cables, just like cutting telegraph cables at the start of the First World War, there was no way to communicate with folks on the other side of the slice.

Researchers such as Vint Cerf and Bob Kahn created hardware, software, and protocols that allowed data to move over networks in a new way: packet switching. In this approach, a message from a computer would be broken into a series of packets (think of an email where every 32 letters would become a new message or packet). These packets would be given a sequence and an address of where the packets should end up. These packets would then be sent to the next computer in the line. That computer was really one of the specialized routing IMPs. The IMP might have two connections to two more IMPs downstream. The IMP could, on a packet-by-packet basis, choose which route to send a packet. If one route was slower, or if one of the IMPs downstream had been destroyed by an atomic bomb, it could send the packet along the faster one.

The packets would bounce through the network independently, finding their own way to the destination machine. The receiving machine would gather the packets together, put them back in order, and display the message.

If some packets were lost, the receiving machine could request that just the lost parts be resent. The system of assigning machines an address on ARPANet was done through a protocol called the Internet Protocol, or IP. The rules for bouncing the packets over the network was called the Transmission Control Protocol (TCP). These two protocols worked together and so are normally referred to as TCP/IP, and are still in use today.

The new approach to digital networking removed the problem of a single point of failure. An atomic bomb over a sensitive phone switch in Kansas no longer meant the west coast was cut off from the east. The packets and communication would just find another path.

ARPANet was so successful that more and more universities connected. As computers became cheaper, organizations such as universities didn't connect simply to access someone else's computing resources, but rather to allow access to their own resources and to support tasks that had little to do with technology or computing. Email was one of the first applications implemented on the network, followed quickly by Usenet,[9] online discussions on topics ranging from computing to Star Trek to music to parenting to, well, pornography.

Soon the institutions signing on to ARPANet had nothing to do with the Defense Department and were using their own money to connect. The ARPANet became just a smaller and smaller part of the Internet. However, the Internet still didn't extend into the world of business until 1990. Like the ARPANet before it, the Internet was primarily for academics and research. There was no commerce, a fact that, as we will see, has ended up with some horrific business and security practices.

FROM EMAIL TO NSA

By 1990 the Internet was a thriving space for the exchange of email, computing resources, and discussion among the academic community. However, the real impact for the knowledge infrastructure went far beyond the advent of email. The packet-switch network that led to a more robust network meant to survive a nuclear war was so successful, it has today completely replaced the old circuit-switched network.

When you make a call on a mobile phone it might seem like you have a direct line or circuit to the receiving end, but in fact your voice is digitized, packetized, and sent across the network. It simply happens so fast, you never notice all the breaking apart and putting back together.

I don't have to spend any time impressing upon you the importance of what this Internet would become. But the story of how a primarily research-driven network moved from the university to everyone's smartphone is in the next chapter. Here I pause to talk about the implications of the nearly invisible and often obscure technology aspects of the Internet , such as TCP/IP and packet switching, have had upon our lives.

You see, the wartime development of the Internet meant not only that communications could be dynamically relayed across the world; it also meant that everything being sent over that network has to be digitized, and packetized. Voice calls that once were audio signals now become digital records on 4G and 5G mobile networks. Images, music, video, and most of the information we seek to share with other people has been transformed into binary data. This, as I'll talk about in chapter 5 and again in chapter 6, means that what we were communicating is now open to be processed, filtered, analyzed, compressed, and then resold at a much larger scale.

In the circuit-switched world of long-distance calls, a conversation disappeared when you hung up the phone unless a significant effort was made to tap and record it. In a packet-switched world, the data that makes up that conversation has to travel over numerous networks and places where it can be copied and collected with ease. Internet technologies created a world in which we have developed a digital trail—the people we call, the places we go, and the apps we use all leave behind digital evidence that is nearly impossible to eliminate by design.

It was the replacement of old phone systems with digital packet-switched networks that led to the creation of the National Security Administration's mass collection of phone records after the terrorist attacks of 9/11. While the NSA had been collecting phone data since its inception in 1952 under the Truman administration, it was limited to intelligence work in foreign communications. That mission was broadened to collect communications, primarily phone communications, from U.S. citizens if part of the call involved foreign agents. The mission was then greatly expanded to include capturing data on nearly all phone communications (the numbers called, for example, not the actual phone conversations) after 9/11.

In 1952, capturing this data involved massive resources to tap wires and create analog phone recordings. By 2011, it was simply a matter of attaching specialized computers to the networks of phone carriers like AT&T. We'll revisit the full impact of the shift when I talk about Edward Snowden in chapter 10.

Of course, all of this would be a moot point if we were only talking about the Internet of the early 1990s. The pre-commerce days of the Internet were far from the graphical experience you are used to today. The interfaces to Internet resources were still primarily text and command lines. To understand how the Internet of "telnet" and "File Transfer Protocol" became the internet of Netflix and Instagram, I have to tell you two stories. The first is how the Internet came to the masses through the World Wide Web, and the second is how the masses pirating music with Napster forced companies to greatly expand the means of tracking people on the net. Both stories are crucial for you to understand how the knowledge infrastructure, and the technologies that connect us to it, are biased and encode human flaws as well as human virtues.

I begin the first story about the World Wide Web in the late 1980s, when a man saw the possibility of the Internet to make a better physics journal.

NOTES

1. "Hiroshima and Nagasaki Missions—Planes & Crews," Atomic Heritage Foundation, April 27, 2016, https://www.atomicheritage.org/history/hiroshima-and-nagasaki-missions-planes-crews.

2. Amy Tikkanen, "*Enola Gay*: United States Aircraft," Encyclopedia Britannica, last modified January 15, 2020, https://www.britannica.com/topic/Enola-Gay.

3. "Hiroshima Log of the *Enola Gay*," Atomic Heritage Foundation, accessed July 14, 2020, https://www.atomicheritage.org/key-documents/hiroshima-log-enola-gay.

4. "Boeing B-29 Superfortress 'Enola Gay,'" Smithsonian National Air and Space Museum, accessed July 14, 2020, https://airandspace.si.edu/collection-objects/boeing-b-29-superfortress-enola-gay.

5. "TNT Equivalent," *Wikipedia*, last modified July 5, 2020, https://en.wikipedia.org/wiki/TNT_equivalent.

6. "The Soviet Atomic Bomb," Atomic Archive, accessed July 21, 2020, http://www.atomicarchive.com/History/coldwar/page03.shtml.

7. "Electromagnetic Pulse," Atomic Archive, accessed July 21, 2020, http://www.atomicarchive.com/Effects/effects21.shtml.

8. "ARPANET—The First Internet," Living Internet, accessed July 14, 2020, https://www.livinginternet.com/i/ii_arpanet.htm.

9. "Usenet," *Wikipedia*, last modified July 8, 2020, https://en.wikipedia.org/wiki/Usenet.

5

The World Wide Web

From CERN to Facial Recognition

Tim Berners-Lee (now Sir Tim Berners-Lee) worked in the library of the CERN research center and was particularly interested in using the Internet to share academic research articles. CERN (in French, Conseil Européen pour la Recherche Nucléaire or the European Council for Nuclear Research) was formed in 1951 out of a United Nations effort to advanced particle physics. Over its lifespan, CERN had become the international leader for exploring matter and the universe at its most basic level. Needless to say, with so much intensive research, the CERN library was very concerned with how physics articles were published.

In the 1980s, much like in physics research some two centuries before, scholars would publish papers in printed journals and include a list of citations to other papers that related to the work. A scholar reading an article would have to go and find these associated documents to fully understand the piece. This involved everything from going to their research library to writing and requesting a copy of a paper from colleagues.

Berners-Lee asked the question, "If we can put an article on the Internet, why can't other researchers as well? Then we can just link to the cited work." In a classic tale of not knowing where a question will take you, the set of software and protocols he developed to create online physics journals was called the World Wide Web. Remember all that talk of http and https in the encryption discussion? That was him.

The World Wide Web didn't create the Internet, and there were (and are) plenty of uses for the Internet that don't involve the web, but Berner-Lee's invention had a massive and quick impact on the knowledge infrastructure at the time.

There were a number of reasons why the web took off. First, it supported a rich graphical experience. Unlike the previous text interfaces, web pages could include images and formatted text. Second, it allowed for hyperlinking, that is, linking images and words to other documents on the web. The third reason is a bit more technical, a general-purpose protocol that was relatively easy to implement in software and could accommodate a growing list of media types (like videos, sound, and eventually apps). The fourth reason is something we will return to in chapter 9; Berners-Lee didn't patent his idea, but gave it away to the world. Each of these reasons would have later, far-reaching impacts. Take, for example, the graphical and formatted text.

By the time the Word Wide Web was developed in 1989, the publishing industry—a key player in the knowledge infrastructure since the time of Gutenberg—had long before adopted computer technology in the production of books and newspapers. It had done so in the late 1960s[1] when computers were still expensive and massive. So rather than giving writers, reporters, editors, and others their own machines, publishers would buy massive computers designed specifically for a given newspaper or type of publishing. Then reporters and writers would work on low-powered word processing terminals. These terminals—think green glowing text on black screens—would put codes around different words and parts of the article. So, there would be a code to say something was a headline, and one for words in italics, and codes for pictures.

The problem was that these codes were unique to each system. If the newspaper wanted to buy a newer, better system, the folks using the word processors would have to learn a whole new set of codes. In 1978 the print industry came together to create a standard set of codes. It was called the Standard Generalized Markup Language (SGML)[2] and became an international standard in 1985.

Now anyone operating a system using SGML knew they could use <H1> to mark a big headline and <i> for things in italics. However, there was a lot of complexity in SGML because it had to cover not just different types of print media (newspapers, books, magazines), but it included different work processes (like how to make a publisher's own codes).

Tim Berners-Lee took SGML, stripped out everything to do with printing or workflow, added special codes to link to other online documents, and called it the HyperText Markup Language or HTML. That "special code to link"? Well, Berners-Lee didn't invent the concept of hypertext and embedding links within documents. That dates back to the work of Ted Nelson in 1963.[3] But Berners-Lee did invent a universal way to point to a resource on the Internet to make hypertext real at scale. And while that may seem techy, it is a set of three letters you probably encounter every day: U, R, L.

Today it may seem as archaic as telegrams, but before 1989 there was no standard way of pointing to an Internet resource, like a page or picture. While machines had names and addresses, resources on the machine did not. So, with earlier systems you would more or less write out directions like recipes: first click here, then here, then here. To solve this problem, Berners-Lee developed the Uniform Resource Locator (URL). Now you could easily point to what machine a document was on, what it was called, and what software (protocol) to use to get it. So http://my.computer.com/directory/article.html meant: Use your web browser (that's the software that "speaks" the HyperText Transfer Protocol or HTTP) to go to the computer named "my.computer.com," then go into the directory named "directory" and grab a page formatted in HTML called "article.html."

The web took the Internet by storm. Universities began putting up home pages. Scholars were putting up articles, and libraries were putting up archives of materials and digital collections. Commerce wanted in.

The World Wide Web with its images and hypertext made navigating the Internet easy. Publishers wanted to publish texts. Newspapers wanted to put their papers online. Stores wanted to show their wares. The only problem was, none of them wanted to do all of that for free. In 1991 the Internet governing body, still located within the U.S. federal government, allowed for-profit organizations to have a presence on the web, and then to conduct business there.[4] The "dot com" was born.

Telephone companies and cable providers saw a new way to sell their home connections and to charge for a new product over the same lines. The internet service provider (ISP) was born. Small regional companies were suddenly competing with international corporations to connect people. As telecommunications became cheaper and bandwidth more plentiful, these same ISPs began to offer connection to the Internet (soon to be ubiquitous enough to lose its capitalization and become simply internet) to people's flip phones. When Apple introduced the iPhone in 2007, the dawn of the modern web everywhere in the world was upon us.*

The web brought the masses to the internet. Yet it wasn't until more and more people connected to the internet that the underlying realities of the internet architecture outlined in the previous chapter become apparent. To understand the true consequences of designing network technology around preparing for war versus preparing for commerce or personal privacy, we have to leave Tim Berner-Lee's world of online physics journals and jump into world of the modern pirate.

FROM CDS TO SPOTIFY

I mentioned before that the ARPANet and the Internet were not built for commerce. They were built by and for scholars, academia, and the military. The military portions on the early internet, known as MILNET, were broken off in 1983 for security reasons,[5] leaving the rest of the network to build the open culture necessary to scientific inquiry. Ideas are to be shared, examined, challenged, and incorporated. That culture can be seen all over TCP/IP. It is the ethos that led to Berners-Lee giving away his invention of the web. However, when commercial interests started coming on the net, they brought a different mindset—but not necessarily a readiness for what they were going to find online. Take Sony.

When Phillips and Sony first released the compact disc in 1982, it was presented as a great advance over previous analog formats like cassette tapes (essentially a long thin strip of cellophane tape coated in rust—easily tangled, easily stretched, easily broken) and vinyl records (scratching, warping, and limited space). Sony launched its CD products by assembling reporters in a room overlooking a huge symphony hall with an orchestra playing classical music. Music filled the room. With great fanfare the Sony representatives pulled back a sheet covering an early CD player and informed the assembled guests that the music they were listening to was coming from the device, not the orchestra behind the glass.

* To be sure, there were plenty of phones that connected to the internet before the iPhone, but the release of the iPhone and its mass-market appeal makes it a convenient point to talk about wide-scale market adoption of mobile internet access.

Compact discs converted analog music into discrete bits, sampling the live music thousands of times per second. These digits were burned onto ultra-thin wafers of aluminum, and the wafers were bonded between two tough sheets of plastic to ensure durability. Not only was the medium more durable, but because it was digital, the new format could incorporate error correction. If the disc was scratched (well, a little scratched), the device reading the disc could find redundant copies of the music data and fill it in with no skip, all without stopping.

The CD was a huge hit with the music industry and its customers alike. The CD turned vinyl records into a medium of collectors and hobbyists, and it killed the cassette tape dead in a few years. Record companies loved the new medium because it was faster and cheaper to produce and ship. It also meant that a lot of music on cassettes and records had to be repurchased to work on CD players.

There was only one little problem with the format. Digital music on CDs wasn't encrypted. As personal computers entered more and more households, and more of those computers had CD players in them, suddenly anyone could copy the music off a purchased CD. Copy it to another disc, or to the computer itself. And what happened when someone put together the fact that digital music was sitting in a computer with the fact that computers were connected to the Internet? In a word: Napster.

Napster was the creation of Shawn Fanning.[6] The idea was pretty simple. A person connected to the internet downloads a small piece of software—Napster—then points the software to a hard drive full of music, and then advertises to the world that it was ready to share. A user with their own copy of Napster could type in the name of a song and send that query out to all of the other Napster installations. A list of songs matching the query would be created, double-click, and suddenly the user is downloading that song from someone else's machine. As more people installed Napster and copied their CD collections, it became easier to find "free" music of just about any genre imaginable.

One Napster innovation that is still with us today was its peer-to-peer design. Just about any computer connected to the internet could share songs. There was no need to have an expensive dedicated machine connected to the internet 24 hours a day. While your personal computer was on the net, it was sharing music. If you turned off the machine (or, say, disconnected your modem on dial-up), then you weren't sharing, and the Napster software would go looking for another copy somewhere else. This may seem like an interesting technical point, but it has profound consequences for the internet, how we think about the web, and how the knowledge infrastructure works.

Peer-to-peer software would spawn a whole host of new systems to allow people who couldn't afford a dedicated server in the network to share their files without an intermediary. Napster users and their descendants would develop ways to share data and change their identifiable addresses online. So even though a person's computer might get a new internet address every time it connected to the internet, proxy services could update the location and allow it to appear to be a stable standing server over time.

These techniques would later be employed by privacy advocates and criminals alike. It allowed creators of content to bypass traditional gatekeepers. Musicians, writers, and illustrators could directly connect with fans and, in the case of Napster, folks hungry for free tunes.

Was sharing on Napster easy? Yes! Legal? Not at all. When you bought a CD, according to the first sale doctrine, you could listen to it. You could re-sell it. You could even loan it to a friend . . . so long as you didn't keep a copy to listen to. But you know what you weren't supposed to do? Make a digital copy and put it on the internet for anyone to copy (still can't). That's piracy (still is).

Napster was HUGE. Some universities had to shut down their internet connections in whole or in part so that professors could teach classes and read their email, because file-sharing traffic ate all the bandwidth. And then the lawsuits started. You see, the Napster software was still built on the idea of an open internet. It didn't do anything to hide which computers were sharing which songs.

Not only were there lists of songs associated with computers, but the very way the internet shipped data around the network exposed who was sharing what. Recall that a message (or song) on the internet is broken into packets, and those packets go all over the net trying to find the receiver. That means that any network used between the sender and the receiver can see what is being shipped. That's why a lot of present-day software encrypts the packets first. Did Napster? Nope.

Record labels gathered huge lists of their intellectual property being shared (pirated) and the addresses of the machines doing the sharing (pirating). The industry then set out to sue the sharers, and that's where the fact that I keep saying they had the "machines" doing the sharing becomes important.

You see, the internet doesn't do people—never has. It was built from the ground up to share data between computers. And remember that many of those foundational pieces of the internet were built around computers that weighed in the tons and might have hundreds of users sharing them. An internet address uniquely identifies a machine, not the person using that machine. Maybe you were sharing songs, or maybe you were using your parents' machine to share songs, or your friend's, or a school machine.

Extensions to TCP/IP and additional protocols make this even more true. Physical computers can use hardware or software to appear as many machines online. This also works the other way with clustering software and virtualization of services such as in cloud computing, where hundreds and thousands of physical machines can appear as a single address on the internet. Google.com is not one massive machine in California—it is thousands of machines located in data centers around the world. Ditto Netflix and Amazon and Facebook and . . . you get the point.

As the record labels went to court, they ran right into the original assumptions that built ARPANet. Courts told the labels they needed more than a machine address; they needed to build a reasonable assurance that only the person being sued was using that machine.

Remember when I talked about the new business model for televisions that watch you watch TV and sell that data? Did you say to yourself, "That should be illegal!" Technically, it is—BUT only if the television company is uniquely targeting an individual viewer. One of the reasons TV manufacturers get past anti-surveillance regulation is the argument that a TV can be watched by anyone who has physical access to it. Most TVs are watched by several people simultaneously, and the companies are doing nothing to identify a single unique individual watching. It is the same argument that universities and those sued (or threatened with a lawsuit) used to avoid tens of thousands of payouts.

Napster was closed down. But then other software popped up to take its place. The only thing that really curbed music piracy (and to be clear: curbed, not eliminated) was Apple's introduction of the iPod and iTunes that made songs easily available for relatively low prices (destroying the record labels' ability to get money for a whole album when the listener just wanted a single hit). Today iTunes and buying music* has been replaced largely by streaming services, like Spotify, that make millions of songs available on demand for a monthly fee or the inconvenience of listening to ads. By 2014 the record industry was worth $15 billion—down from a high of roughly $38 billion in 1999.[7]

FROM CHATBOTS TO BROWSER FINGERPRINTS

This fact that internet infrastructure is not about people and does not, at the protocol level, carry individual identity information has been used in pedophilia cases, computer hacking prosecution, even divorce cases. It is also the reason that services like Facebook and Twitter now have as many chatbots and software systems used to spread propaganda as they do real human users. It is the same reason that you often have to work pretty damn hard to prove you are a human online. Passwords, first pet's name, mother's maiden name, click all the images with a car in it, conduct a simple math problem—all of these are ways that entities on the internet try to identify you uniquely. If you've had to copy a texted code off your phone onto a web page, blame ARPANet.

That said, it's not like folks haven't been trying to uniquely identify you without passwords (or faces or fingerprints). However, this is not always in your favor. Identity thieves work hard to figure out how to impersonate you by copying images from social media profile pictures, or intercepting credit card numbers or passwords, without you knowing. That way they can provide the password, click on all the car pictures, or let folks know that Fluffy was your first pet. Now maybe you know why there are so many Facebook quizzes that try to tell you if you're a cat or dog person.

Most modern web browsers come with some sort of private or incognito mode. Even without these, most also offer some sort of setting to tell websites not to track you. Yet, even with these features, that handbag or watch you were shopping for on Amazon keeps showing up in advertising on Facebook and the other sites you visit. How? Well, one of the most used tracking techniques can be blamed on how Tim Berners-Lee designed the web, and one can be blamed on publishers trying to make your web pages more stylish.

When you write a paper or a memo on a traditional word processing program like Microsoft Word and you include an image or two, you simply

* Remember, however, the discussion of DRM and not really "owning" digital music in chapter 2.

paste the images into the document. From that point forward, as far as you are concerned, the images are part of that single file stored on your hard drive. HTML, the primary file type for the web, doesn't work that way. It stores text and all of those formatting tags (like <H1> for headings), but it can't store complex information (technically binary files) like images. Clever Berners-Lee had a ready solution. The same URL standard that pointed to other HTML files could also point to image files. So to add a picture to your web page, you insert an HTML tag like this: <img src=URL>. That tag tells your web browser, ad, phone—whatever—to go to the given URL, download the image, and put it into the document. While the viewer of the page sees it as a single file, it is in fact two. That URL could point to an image on the same computer, or anywhere else on the internet.

Let's say someone wants to track you surfing the web without you knowing it. First, you need to load a page they put up. In that page is a pointer to a single-pixel invisible image. You never know it's there, but when you called up the page, your online stalker (or more likely advertiser) now has the address of your machine and what page you were looking at. You continue to surf the web and come to another page on another server that, unbeknownst to you, is working with stalker/advertiser. As you load this new web page, a single 1 x 1 pixel invisible image is loaded . . . the same one as before. Now your stalker knows the address of your machine that looked at the first page is now looking at another page. If that first page was product information on a new backpack, then the two sites working together can place an ad for that same item in the new site. Furthermore, wherever you go and load that invisible image file, your stalker can now keep track of where you went, what you looked at, and how long it took you. These are called bugs . . . not software errors, but bugs like listening devices.

Browsers are on to this and use different techniques to counter them. This includes changing your IP address as you surf. However, if you don't change browsers or the computer you are surfing with, advertisers and stalkers use another technique made possible because the original HTML language made pages that looked simple and ugly.

As publishers and the media sought to use the web, they were frustrated by just how limited a design canvas they had to work with. Bold, headers, and italics might work for physics papers, but not for magazines or graphics-intensive textbooks. Publishers needed more control over how things looked, and that meant they needed to know what they had to work with. Different early web browsers, for example, supported different HTML tags or made the tags look different. Not every computer had the same fonts loaded. A designer could make a page look beautiful using the Helvetica font loaded in Netscape, but your machine only had Ariel installed and used Internet Explorer, and the page composition went haywire.

So, the web standards body that now included publishers began requiring your browser to send a lot of information to a web server when requesting a page. Instead of simply asking for a page, your browser would provide the server information such as which browser software you were using, a list of the fonts installed on your machine, the dimensions of the screen you were using, the maker of the machine everything was running on, and so on. Designers could use this information to reformat pages on the fly for optimal layout.

However, people in the tracking business realized that all this information combined formed a unique fingerprint of your machine. No need to load an image, no need to see the same address—they could just look for your fingerprint. If you'd like to see your fingerprint, take a look at the Electronic Frontier Foundation's Panopticlick.[8] Figure 5.1 shows the results for the machine I am working on while writing this.

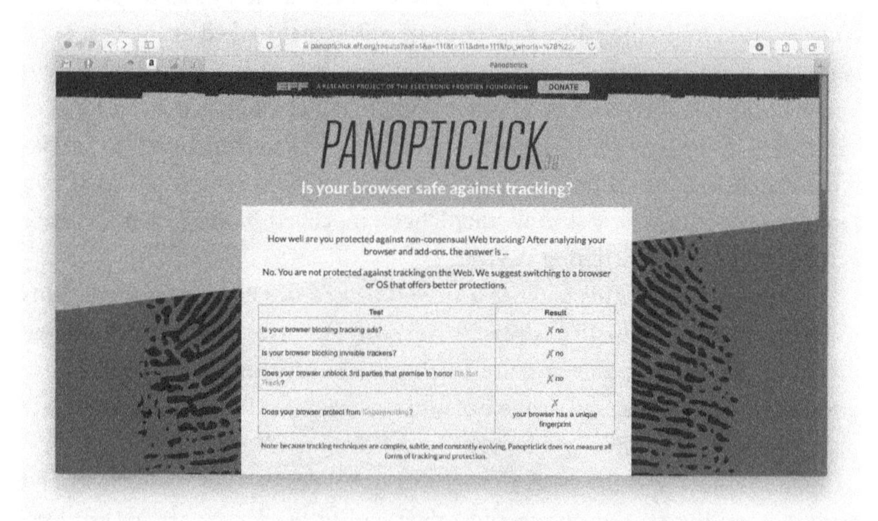

FIGURE 5.1
A basic fingerprint of a browser. Here's a hint: this is not a good result. *"Panopticlick," Electronic Frontier Foundation, accessed July 21, 2020, https://panopticlick.eff.org*

The line "Is your browser blocking invisible trackers?" is the image bug I just wrote about. But what does my fingerprint look like, and how unique is it? See figure 5.2.

By blocking bugs or using my browser's private mode I may feel like I'm being more private on the web, but with the other data I provide or, rather, often unknowingly, my software provides, I am not. Worse still, all of those private modes and "do not tracks" create a false sense of safety for internet use that can lead to us exposing more sensitive information than we might

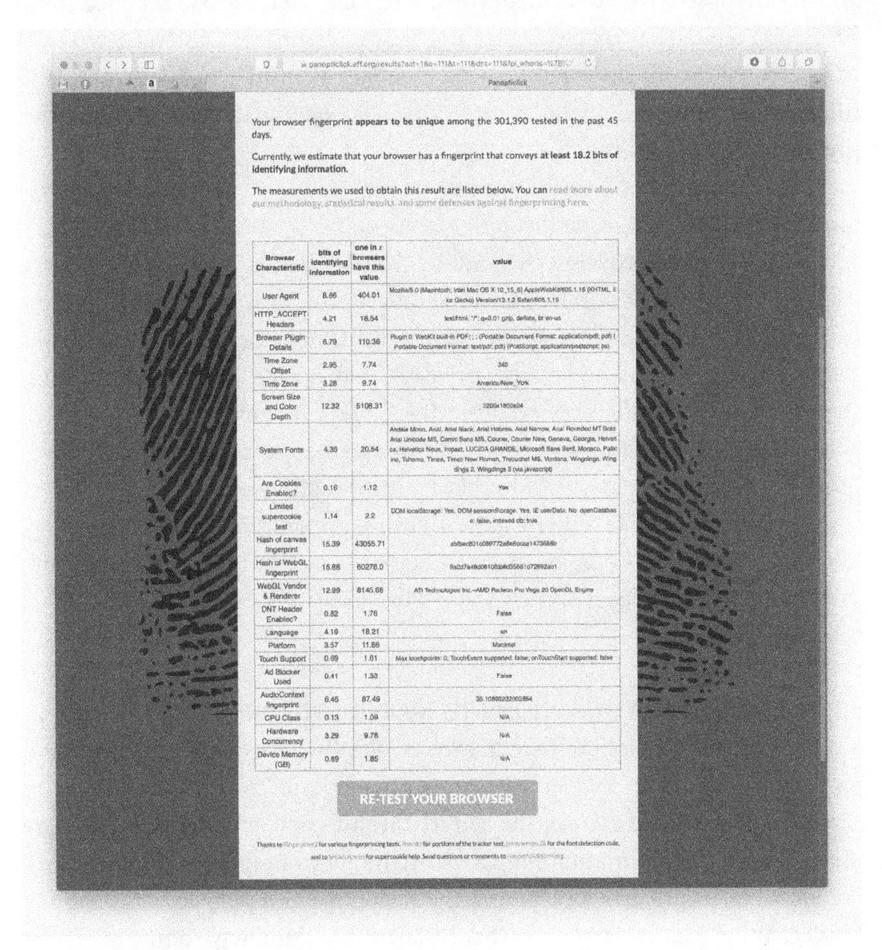

Panopticlick

Your browser fingerprint **appears to be unique** among the 301,390 tested in the past 45 days.

Currently, we estimate that your browser has a fingerprint that conveys **at least 18.2 bits of identifying information.**

The measurements we used to obtain this result are listed below. You can read more about our methodology, statistical results, and some defenses against fingerprinting here.

Browser Characteristic	bits of identifying information	one in *x* browsers have this value	value
User Agent	8.66	404.01	Mozilla/5.0 (Macintosh; Intel Mac OS X 10_15_6) AppleWebKit/605.1.15 (KHTML, like Gecko) Version/13.1.2 Safari/605.1.15
HTTP_ACCEPT Headers	4.21	18.54	text/html, */*; q=0.01 gzip, deflate, br en-us
Browser Plugin Details	6.79	110.36	Plugin 0: WebKit built-in PDF; ; ; (Portable Document Format; application/pdf; pdf) (Portable Document Format; text/pdf; pdf) (PostScript; application/postscript; ps).
Time Zone Offset	2.95	7.74	240
Time Zone	3.28	9.74	America/New_York
Screen Size and Color Depth	12.32	5108.31	3200x1800x24
System Fonts	4.36	20.54	Andale Mono, Arial, Arial Black, Arial Hebrew, Arial Narrow, Arial Rounded MT Bold, Arial Unicode MS, Comic Sans MS, Courier, Courier New, Geneva, Georgia, Helvetica, Helvetica Neue, Impact, LUCIDA GRANDE, Microsoft Sans Serif, Monaco, Palatino, Tahoma, Times, Times New Roman, Trebuchet MS, Verdana, Wingdings, Wingdings 2, Wingdings 3 (via javascript)
Are Cookies Enabled?	0.16	1.12	Yes
Limited supercookie test	1.14	2.2	DOM localStorage: Yes, DOM sessionStorage: Yes, IE userData: No, openDatabase: false, indexed db: true
Hash of canvas fingerprint	15.39	43055.71	[illegible]
Hash of WebGL fingerprint	15.88	60278.0	[illegible]
WebGL Vendor & Renderer	12.99	8145.66	ATI Technologies Inc.~AMD Radeon Pro Vega 20 OpenGL Engine
DNT Header Enabled?	0.82	1.76	False
Language	4.19	18.21	en
Platform	3.57	11.86	MacIntel
Touch Support	0.69	1.61	Max touchpoints: 0; TouchEvent supported: false; onTouchStart supported: false
Ad Blocker Used	0.41	1.33	False
AudioContext fingerprint	6.45	87.49	35.10893232002864
CPU Class	0.13	1.09	N/A
Hardware Concurrency	3.29	9.78	N/A
Device Memory (GB)	0.89	1.85	N/A

RE-TEST YOUR BROWSER

FIGURE 5.2

An example of a browser fingerprint. *"Panopticlick," Electronic Frontier Foundation, accessed July 21, 2020, https://panopticlick.eff.org*

otherwise do. Who in the world would use a work computer to access pornography, you might ask? Well, someone who thinks that work can't track usage in incognito mode.

That last point—knowing what your software is doing—is not a minor point. As individuals and communities become increasingly dependent on apps and digital transactions in the knowledge infrastructure, the fact that more and more people are depending on something they know less and less about matters. I've already mentioned web browsers and TVs, but even when you are supposed to be informed, you are leaking data. In 2015 the Electronic Frontier Foundation reported that the U.S. government was "quietly sending personal health information to a number of third party websites. The information being sent includes one's zip code, income level, smoking status, pregnancy status and more."[9]

The offending site? HealthCare.gov—a required tool for people to sign up for healthcare not available through employers.

In 2014 it was discovered that Adobe Acrobat Reader, a piece of software used by thousands of libraries to distribute ebooks, was sending reading information back to Adobe.[10] What is very disturbing about the incident was that libraries—bastions of reader privacy—had no idea, and were assuring their members that the app wouldn't track them. Yet every book borrowed and every page turned was transmitted, unencrypted, and sent not back to the library, but back to a corporation that made the software.

Through things like super cookies, browser extensions, Wi-Fi tracking, and much, much, more, advertisers and others can follow you. And these are only the techniques legal and ethical parties use (well, maybe not in the last two examples).

FROM FOLLOWING YOUR PHONE TO BIASED SYSTEMS

With this ability to follow people around the Internet with such precision and invisibility, is it any wonder governments and companies alike want more people online? Across the globe, the internet is ubiquitous. We are always online. Even in poor countries, technology companies like Facebook are working to provide free internet connections and very low-cost phone service. Why? Humanitarian reasons? Maybe, but definitely so that everyone can generate data that can be collected, analyzed, and ultimately monetized. Corporations are extending the digital nature of the knowledge infrastructure to increase profits. Corporations and governments alike are actively shaping the policies and technologies of the infrastructure to gain wider access to the personal data of more people.

And the tracking is moving from the web to the phone. Whereas IP addresses only identify machines, phones are a different matter. While technically they too can be shared, in reality, they are one-to-one machines. Just as web browsers share a lot of data, so do apps on phones. Think of all the ways a phone uniquely identifies you and what you are doing. You have a unique phone number so that people can call you. Your phone can give you driving and walking directions, so it knows where you are. Phones reorient the screen from landscape to portrait as you turn them or measure movement for games, so the phone knows how you are moving. Researchers have shown these capabilities can be used for gait identification—identifying you by how you walk. Digital assistants like Siri or Hey Google can listen for a keyword to tell you the weather, read a text message, or perform some other function, hands-free. This, of course, means your phone is always listening to you. Phones are tracking you and creating digital trails of your life, not just when the screen is on, but so long as they are drawing power.

And so, we have come to another crucial component of understanding the knowledge infrastructure: universal connectivity. Just as with data and media, connectivity is a concept that is often discussed in near-utopian terms, but can also be seen as a societal cost, even disease.

We read articles on the danger of always being on our phones, on our phones. We worry about how early exposure to technology affects child brain development while we watch our infants through digital video cameras that store the previous 12 hours of their lives in the cloud. Google Photos, Facebook, and Apple's Photos create automatic movies for our birthdays, while storing the location of our schools, friends, and houses of worship.

The knowledge infrastructure that used to be analog and "opt in"—that is, you had to seek out sources of information in newspapers or libraries—has become digital and ubiquitous. It has become essential to modern life, and now it has a toll booth that collects your personal data as a token of passage. By the way, a toll not just for consenting adults, but for our children as well.

South Carolina is seeking to improve education, and has bet on technology as a way to do it. The state heavily subsidizes internet connectivity to the poorest schools in the state. It funds one-to-one programs in school districts where each student will have their own tablet or laptop. These investments are then used to deliver digital textbooks, online homework, and testing.

The only problem with online educational materials is that while students may be online in schools, they may not be at home. In the most rural parts of the Palmetto State, kids don't have access at home. It is either unavailable, very slow, and/or very expensive.

To bridge this digital divide, the state is supporting schools loaning out Wi-Fi hotspots. These are handheld cellular modems that connect to an internet service provider and then set up a small Wi-Fi signal to which students can connect their tablets or laptops. Other schools are taking advantage of the fact that some students live in such remote parts of the state, they may be on the school bus for an hour or two getting to school and home again. These districts are wiring the busses to be hotspots for these long bus rides.

All of this extended education infrastructure is about connecting devices such as iPads and Chromebooks to the net. Devices and services like Google's G Suite for educators are uniquely identifying a child as well as their academic performance and confidential communications with teachers and counselors.

But at least the iPads and Chromebooks are visible means of tracking. Increasingly, the technological devices tracking us are invisible. Facial recognition, another technology developed with military funds, uses images to identify individuals. It is often used in private and public settings for security. The idea is to find bad actors before they, well, act badly. In private settings, you are supposed to be informed that cameras are in use. However, in public settings, you are assumed to have no privacy, you choose to be there, and therefore no such indication is required. But what about doorbells?

Video doorbells are a growing class of internet devices that didn't exist either when the internet started or when people wrote laws on things like cameras in the public sphere. Many states prohibit home recording of public streets. Yet video doorbells, which are supposed to help a private individual identify a person on their front porch, often take in the streets in front of the houses. That alone is not necessarily a bad thing, but remember we are talking about pervasive networks—these doorbells are connected.

Hardware/service combinations like the Ring doorbell record the porch and the street, and then send that data into the network. There homeowners can answer the door or review the footage from anywhere in the world . . .

but so can others. The *Washington Post* reports on how Ring works with law enforcement agencies to access this data:

> The partnerships let police request the video recorded by homeowners' cameras within a specific time and area, helping officers see footage from the company's millions of Internet-connected cameras installed nationwide, the company said. Officers don't receive ongoing or live-video access, and homeowners can decline the requests, which Ring sends via email thanking them for "making your neighborhood a safer place."[11]

Did Ring doorbell owners realize they were becoming part of a neighborhood watch when they bought that device (or received it as a gift)? Did they take that into consideration when they aimed the camera? Today we—the citizens—have put in place more surveillance capability than the British Empire could have dreamed of when the CS *Alert* was sent to cut Germany's submarine cables. And a lot of us have done it unknowingly by simply signing on to Facebook or Twitter.

A company called Trustwave released a piece of facial recognition software that scrubs through sites like Facebook and Twitter, enabling anyone with access to the software to track people from site to site.[12] Companies like Clearview[13] are using these kinds of tools to build massive databases of people linked to facial recognition software for both public surveillance and tracking online behaviors. Companies like Clearview, but also companies like Amazon do this.

"Why," you might say, "do I care if an online book seller sells facial recognition software?" The short answer is that Amazon outgrew the title of bookseller a decade ago. Remember that Amazon produces and sells Echo and Alexa software and devices that literally listen to you all the time, waiting for a keyword to activate. These sound files have been subpoenaed in court. This is an obvious Amazon product. But what about the services you use that depend on Amazon?

One of the profitable parts of Amazon that few people outside of the tech industry pay attention to is Amazon Web Services (AWS). This is an enormous set of computing tools and storage built on a worldwide network of data centers like those owned by Google. You may not have heard of AWS, but you've probably heard of the companies that use it: Netflix, LinkedIn, Facebook, the BBC, ESPN, Adobe, and Twitter.[14] You may also be interested to know that up until the racial protests of 2020, Amazon sold a facial recognition software service to law enforcement and the military.[15]

Now to be very clear (and hopefully avoid a lawsuit), Amazon is not using all the data on your face, your likes, your tweets, your movie viewing, your book reading, and your sports preferences across these services to power their software. Each service uses Amazon's hardware and software but maintains its own private data and use policies. But all of that data is now in one place and dependent on one company. That does have implications for your data, and as we will see in the "Media" section of this book, a lot of implications for how media works today.

This is where one might come to the conclusion that network technologies, and indeed technology in general, can be either good or bad. Used to track you, but used to keep you safe. The internet, the laptop, the phone are all neutral pieces of functionality, and people put these to good or bad use, you might conclude. The internet allows us to learn from others across the globe and track us while we have these interactions. Our phones are always connected, but it also means we are always sharing the details of our lives. Technology can either help or hurt, but it is the uses, not the technology itself, that determines bias. Yeah, don't come to that conclusion—it's wrong.

The knowledge infrastructure we use is infused with bias. From giving preferential access to certain types of people (overwhelmingly white, male, and monied), to building in technologies enabling commercial tracking, to a policy and legal framework that benefits intellectual property holders over creators and consumers alike, the infrastructure reflects its creators and owners. As we will see again and again and again, our knowledge infrastructure has baked in winners and losers.

The twist is, it always will, no matter how hard we seek an objective system. That is because knowledge is uniquely human, and humans cannot be unbiased. Therefore, any system we build to disseminate and facilitate knowledge will be biased. The point, as we will explore in depth in the "Society" section of this book, is not to seek an impossible standard for "unbiased" or "objective," but rather to make the biases obvious, examined, and agreed upon.

Just as the *Enola Gay* launched a society-wide conversation about the use of nuclear weapons in the military infrastructure, the launch of ARPANet, the development of the internet, and the rise of the web should have begun a conversation about the role of the individual in the knowledge infrastructure.

Instead, for much of the history discussed here, the focus was on institutions, and on technological capabilities. We spent thousands of hours at conferences and millions of pages in books and journals discussing the impact of the internet on libraries, and publishers, and newspapers, and television, and governments, but only a fraction of that effort on citizens and individuals. Just as the internet was built to handle machines, not people, our copyright laws are shaped by industry, creating a system of infractions and enforcement over creation and sharing. Not just by the courts, but in how we teach plagiarism to our college students and information literacy to our grade schoolers.

It is only recently, with major data exposure scandals at Facebook and Ashley Madison and documented evidence of election manipulation by Russia, that we have begun to see the true cost of our present knowledge infrastructure. If there is one point you take away from this book, it is this: How our infrastructure works—which groups it preferences and what behaviors it dissuades—is our choice. When it was in the national interest in World War I to transform the radio waves from an owned monopoly to a common resource, policy action across borders made it happen. When a military research resource was found to be useful by scholars and students alike, it was transformed into first a civilian operation, then opened up to international governance. Now we must see how all of this infrastructure benefits the individual, provides agency to people rather than institutions, and is put to the common good.

We have now covered major technologies that shape our daily lives: encryption, massive-scale computing, the internet, and the world wide web. All of these technologies have certainly changed the knowledge infrastructure, giving us ready access to information from across the globe in milliseconds. But all of these technologies are built on the more diffuse concept of data and are guided by national strategy. To finish up the understanding of the data portions of the knowledge infrastructure, we must further develop our understanding of data, where it came from, and where it takes us. And so, we jump back in time to the work of a mathematician with a measurement problem.

NOTES

1. Thomas Haigh, "Remembering the Office of the Future: The Origins of Word Processing and Office Automation," *IEEE Annals of the History of Computing* 28, no. 4 (2006): 6–31, http://doi.org/10.1109/MAHC.2006.70.

2. "A Brief History of the Development of SGML," SGML Source, June 11, 1990, http://www.sgmlsource.com/history/sgmlhist.htm.

3. "Hypertext," *Wikipedia*, last modified July 17, 2020, https://en.wikipedia.org/wiki/Hypertext.

4. "History of Ecommerce," Commerce-Land, accessed July 21, 2020, https://www.ecommerce-land.com/history_ecommerce.html.

5. Conrad Johnson and Brian Donnelly, "A Brief History of the World Wide Web and the Internet" (prepared for the Columbia Legal Theory Workshop, New York, October 24, 2003), http://www2.law.columbia.edu/donnelly/lda/ih/techprofx3.html.

6. Tom Lamont, "Napster: The Day the Music Was Set Free," *Guardian,* February 23, 2013, https://www.theguardian.com/music/2013/feb/24/napster-music-free-file-sharing.

7. Eamonn Forde, "Oversharing: How Napster Nearly Killed the Music Industry," *Guardian*, May 31, 2019, https://www.theguardian.com/music/2019/may/31/napster-twenty-years-music-revolution.

8. "Panopticlick," Electronic Frontier Foundation, accessed July 21, 2020, https://panopticlick.eff.org.

9. Cooper Quintin, "Healthcare.gov Sends Personal Data to Dozens of Tracking Websites," Electronic Frontier Foundation, January 20, 2015, https://www.eff.org/deeplinks/2015/01/healthcare.gov-sends-personal-data.

10. Sean Gallagher, "Adobe's e-Book Reader Sends Your Reading Logs Back to Adobe—In Plain Text" [Updated], Ars Technica, October 7, 2014, https://arstechnica.com/information-technology/2014/10/adobes-e-book-reader-sends-your-reading-logs-back-to-adobe-in-plain-text.

11. Drew Harwell, "Doorbell-Camera Firm Ring Has Partnered with 400 Police Forces, Extending Surveillance Concerns," *Washington Post*, August 28, 2019, https://www.washingtonpost.com/technology/2019/08/28/doorbell-camera-firm-ring-has-partnered-with-police-forces-extending-surveillance-reach/?arc404=true.

12. Russell Brandom, "New Facial Recognition Tool Tracks Targets Across Different Social Networks." The Verge, August 8, 2018, https://www.theverge.com/2018/8/8/17663640/socialmapper-facial-recognition-open-source-intelligence.

13. https://clearview.ai is the company's home page, but you might be interested in this article on the service and the controversy it started: Kashmir Hill, "The Secretive

Company That Might End Privacy as We Know It," *New York Times*, last modified February 10, 2020, https://www.nytimes.com/2020/01/18/technology/clearview-pri vacy-facial-recognition.html.

14. Ben Saunders, "Who's Using Amazon Web Services?" [2020 update], Contino, January 28, 2020, https://www.contino.io/insights/whos-using-aws.

15. Kelly Tyko, "Amazon Shareholders Reject Banning Sale of Facial Recognition Software to Law Enforcement," *USA Today*, Last modified May 23, 2019, https://www.usatoday.com/story/tech/2019/05/22/amazon-facial-recognition-shareholders -reject-banning-software-sales/3770927002.

6

Dataism

From Reducing Noise to Reducing People

It is at this point we revisit the work of Bell Labs, the same Bell Labs that developed the transistor. The lab drew the smartest minds from academia and beyond. It also did work for the U.S. government. In fact, Alan Turing spent time there after the Second World War. While he was visiting, he interacted with another mathematical genius, Claude Shannon.

Shannon, like the CS *Alert*, is another fascinating character in our story that serves as a nexus of historical events. Aside from his interactions with Turing, Shannon worked on Vannevar Bush's differential analyzer at MIT. During World War II Shannon worked on both cryptography—the work that would bring him into conversations with Turing—but also on fire-control systems that sought to guide anti-aircraft guns through automatic calculations of ballistic trajectories, the same kind of work that led to the development of the digital computer. He also virtually invented the conception of data that we use today.

In his work on cryptography and secure telephones (this was Bell Labs, after all), Shannon kept running into a problem. The problem was in measurement. Telephone systems to this point were purely analog, and a huge engineering problem in connecting analog phone calls over long distances is interference. The phone company had to continually boost the electrical signal traveling from the source to the recipient. However, there would always be interference, and when a signal is boosted, the interference is boosted as well. Soon enough the minor hisses and clicks caused by stray electrical signals on a call could overcome the signal, and the listener couldn't understand what was being said.

The problem is worse when you are trying to encrypt the signal to keep it secure. Here you were deliberately distorting the signal at the outset. The problem on the receiving end is to not only decrypt the message but determine what was "message," what was encryption, and what was just noise. Shannon was a precise mathematician, and he needed ways to quantify things like noise and signal. He wanted to figure out ways to determine the bare minimum amount of signal one needed to communicate. All of this work was distilled as information theory.[1] It used ideas like information entropy, signal-to-noise ratios, and quantification of information to describe how information could be stored, transmitted, and encoded. It also introduced the concept of the bit.

The bit was the smallest discernable amount of information in a system. In a digital computer, for example, a bit is either a 1 or a 0. You could string bits together to make larger units like, say, a letter. A letter in a computer is represented by 8 bits—more commonly referred to as a byte. For example, the letter A in binary is 01000001. B is 01000010. Note that capitalization matters. Lowercase b is 01100010. Storing text is so fundamental to computers that the smallest addressable space in a computer's memory is a single byte, not the smaller bit. This is why computer storage is measured in bytes.

On the other hand, the smallest amount of information that can be sent over a communications channel, say the internet or your mobile phone connection, is a single bit. That's why your internet speed is measured in bits. So, your computer's hard drive might store 512 giga*bytes*, but it might have a 100 mega*bit* per second connection through an internet service provider.

This may seem extremely basic, and perhaps something you were forced to learn in computer classes in school. However, at the outset, it unlocked some amazing capabilities. For example, not only did it allow for once analog signals to be digitized (leading to inventions like the compact disc), digital signals could be securely encrypted (unlike the compact disc). It also allowed for data compression. Data compression is as essential and invisible to today's knowledge infrastructure as encryption.

Data compression is how Netflix can send full-length movies to your mobile phone in high definition. It is also how Netflix streamed 350 million hours of video to their subscribers on a single day, January 7, 2018.[2] The movies you watch on YouTube are compressed; the pictures you view on web pages are compressed. Even the voices you hear on your phone are compressed. Compressed using Claude Shannon's work, and used today to make content small enough to fit on glass fibers crisscrossing the globe or bouncing off satellites flying 22,245 miles above the Earth.

It is important to note that compression and encryption share a common weakness. If you lose access to the algorithms or keys used to manipulate the data, you can lose the data forever. And while you might think the answer is to, well, never lose those things, it is not always that simple. For example, many of the methods for encrypting and compressing are proprietary. That is, someone owns the method. What if that person (or more likely, institution) decides not to share that method any longer? Well, you've lost the keys and, most likely, the data. Mountains of documents, government reports, and home videos have been lost to history because companies have gone out of business or updated software without backward capability. This fact is one argument for a vibrant and public sphere.

As more and more information and crucial documents in society become digital and, oftentimes, compressed and/or encrypted, preserving them over time becomes more and more complex. Whereas an archeologist can decipher tablets and scrolls written thousands of years ago, computer scientists are often unable to retrieve the contents of files stored on outdated media like Zip drives or saved in files created by early word processors such as Scripsit. We will return to this issue of preservation in the knowledge infrastructure in chapter 8. For now, the important takeaway is that our modern conception of data was as a way to compress communications—to simplify things to their bare essence. And in doing so, we often strip away vital information—information that allows us to find meaning in the future. To find meaning in the fact, for example, that after World War II, there were Nazis living in Florida.

FROM PAPERCLIPS TO RESEARCH UNIVERSITIES

Shannon's primary funder in his work on targeting systems, encryption, and information theory was the U.S. government and its military. It is hard to argue with the impact that government funding has had on telecommunications. In fact, if World War II taught governments and militaries anything, it was the value of investing in basic research. The Enigma encryption was broken in large part by college professors working for the military. Ditto the first digital computers and the atomic bombs dropped on Japan. Bell Labs would mix government and corporate funding to go on to develop solar cells, lasers, and communication satellites. BBN, the engineers behind ARPANet and the internet, was a basic research company. These think tanks often blurred the boundaries between corporation, government, and academia.

This lesson was also learned by Germany in World War I and leading up to and during the World War II. German military science developed paratroopers, helicopters, and jet engines;[3] very nearly developed atomic weapons; and deployed the first rockets in the form of V2 missiles.

The scientific talent of the Nazi war machine did not go unnoticed by either the Americans or the Soviets. At the end of the European conflict in the Second World War Soviet troops advanced into Germany from the east as the United States and its allies advanced from the west. In the process they captured German scientists, engineers, technicians, and reams of scientific documents. These were sent back covertly to the United States and the Soviet Union as each drew closer to Berlin. The operation in the United States was called Paperclip, and among other things led directly to the U.S. space program. The Soviet program was called Operation Osoaviakhim, and it relocated more than 6,000 people—scientists and their family members—to the Soviet Block.[4] It too would result in a space program, one that would put Sputnik, the first artificial satellite, into space.

Aside from the scientists themselves, the United States scooped up literally tons of books, foreign documents, and German propaganda. This effort actually began at the outset of the war when the Office of Strategic Services (OSS), one of the first dedicated U.S. intelligence agencies, teamed with the Library of Congress to recruit librarians and bibliophiles to acquire, catalog, and analyze foreign books and newspapers. These efforts led to both the identification of information as an asset and the rise of information science to study how such information could be sought, organized, and disseminated beyond the context of libraries* and outside the confines of a book or printed document.

What happened with the German scientists and technical knowledge shapes how we think about the knowledge infrastructure, particularly the public and open aspects of it. Are we currently continuing to build upon a successful open

* Just go read the brilliant *Information Hunters: When Librarians, Soldiers, and Spies Banded Together in World War II Europe* by Kathy Peiss (New York: Oxford University Press, 2020).

architecture of innovation and discovery? Or are we busily replicating closed systems of advancements guarded behind patents and corporate secrecy?

In the Soviet Union, the German scientists served as a massive infusion of expertise. Upon the Osoaviakhim foundation the Soviets built a centralized scientific and technical information system. Rather than work with universities or companies (that didn't have a major presence under the communist system), the Soviets built a series of specialized research centers reporting directly to the Soviet state and focusing on physics, nuclear research, medicine, engineering, and other fields. The documents and information produced at these centers became part of VINITI (an acronym for the Soviet Vserossisky Institut Nauchnoi I Tekhnicheskoi Informatsii or "All-Russian Institute for Scientific and Technical Information"). Access to this information was severely limited. Russian scientists were not permitted to collaborate with foreign scientists, for example.

> VINITI formally launched in 1955, and became a massive information operation.
>
> According to a brochure published in 2002, VINITI today is (1) a team of highly skilled professionals processing more than one million scientific publications annually, (2) the largest on-line data bank in Russia, which contains more than 25 million documents, (3) the Referativnyi Zhurnal (the Abstract Journal), which is distributed, subscribed to, and read by scientists in sixty countries, (4) more than 330 titles of publications, which cover all fields of basic and applied sciences, (5) more than 240 databases on science and technology, economics, and medicine, (6) powerful retrieval systems, and (7) a wide range of services.[5]

And yet, by 2002, VINITI supported a failing technical and scientific infrastructure.[6] The early successes with rocketry and nuclear power had begun to fade, and in terms of information technology neither the Soviet Union nor Russia were ever seen as strong players.

In the United States, the scientific and technical information system took a radically different approach. Rather than invest in new centralized academies and scientific dissemination systems, the United States distributed the function. While there were important governmental clearinghouses of information, such as the Defense Technical Information Center (DTIC),[7] much of the information was shared across universities and research centers, driven by new advances in information science born in the war and a fundamental redefinition of information from a thing (book or document) to that which could be extracted from a thing (an idea or understanding). Where the government did fund information centers, they were often hosted and operated by universities. New agencies were formed, such as the National Aeronautics and Space Administration (NASA) and the National Science Foundation (NSF) to encourage open research and exchange of ideas. National libraries in medicine, agriculture, education, and transportation were created to both gather and disseminate the latest thinking in their areas.

The AskERIC project I mentioned earlier? That was part of Education Resources Information Center (ERIC), a series of subject-specific clearinghouses located around the country to support teachers, education scholars, and policy makers. The clearinghouses were funded by the U.S. Department of Education but housed in universities and research centers across the country. The U.S. Department of Defense ran a similar system known as Information Analysis Centers for engineering and open military information. The U.S. system didn't just fund research; it funded the dissemination of that research. Grants to universities and research centers regularly included funds for publishing results and hosting gatherings, briefings, and conferences. It also funded graduate students to make up the next generation of researchers.

The results? Well, look at the running list of innovations I've covered so far: the internet, digital computers, rockets, satellites, mobile phones, Wi-Fi, the transistor, and we could keep going on for multiple volumes. The U.S. model was replicated and connected abroad to the United Kingdom, Canada, West Germany, Korea, Japan, and most of the non-Soviet world.

The reason I present both the Soviet system and the American system is not out of a sort of national pride, but to demonstrate that choices have consequences. VINITI didn't just organically happen—it was a government choice, followed by funding, laws, policies, and very real consequences. Likewise, the U.S. system was the result of people making choices—several of those key individuals mentioned in this text. It also has had some VERY real consequences for today's knowledge infrastructure. Many we have already talked about, such as omnipresent encryption and networks supporting an infrastructure that has transformed the largest corporations in the world into surveillance organizations. But some consequences were not anticipated when the investing began.

FROM COEDS TO COEFFICIENTS

After World War II, the government continued to pump billions of research dollars into American universities. To get as many of these dollars as possible, universities increased the credentials of the faculty, now making a doctorate the primary degree of researchers and thus increasing the pay of academics. Administrators were hired to locate, seek, and manage funded research. Universities hired lobbyists to push for even more funding. Graduate education greatly expanded, and the focus on undergraduate teaching within research universities now had to compete with scholarship and increased service obligations of faculty.

The presence of federal funding allowed states that had been the primary funders of public higher education to lower their contributions. States have cut annual funding of higher education by $9 billion over the last 10 years.[8] Large research universities like the University of Michigan receive 14.1 percent of their funding from the State of Michigan, on par with the 12 percent that comes from sponsored research. This withdrawal of state funds has meant that universities have increased tuition (73.4 percent of UM's funding).[9] The result? A nation of citizens that owe more in student loans than credit card debt.[10] In 2019, *U.S. News* found that over the past 20 years:

- The average tuition and fees at private National Universities have jumped 154%.
- Out-of-state tuition and fees at public National Universities have risen 181%.
- In-state tuition and fees at public National Universities have grown the most, increasing 221%.[11]

These numbers, and the forces of ubiquitous networks, data, and computing, are also changing the enterprise of higher education. With such large national investments, there have been understandable calls for accountability and the adoption of efficiencies, most often (and often in a misguided way) attributed to business.

FROM THE DATA TO DATAISM

Higher education is a fantastic example of the rise of *dataism*. Dataism is a term first coined by David Brooks in 2013[12] and explored by Yuval Harari in his book, *Sapiens*,[13] and more deeply in his follow-up, *Homo Deus*.[14] It is an ideology that models the world as a constant stream of information; governments, businesses, society, and people in general are processors of data in that stream. Dataism can be seen in the advent of data science and analytics that replace traditional scientific methodologies with large-scale data gathering, statistical manipulation, and the use of algorithms such as deep learning that we discussed around artificial intelligence in chapter 3.

The issue with dataism is that it tends to be a breathless belief that all insight can be found in data collection. In the sphere of scholarship, it tends to either minimize or outright reject the scientific process of hypothesis testing. In effect, if not in explicit declaration, dataism believes that sufficient quantifiable data collected in the daily workings of digital life can yield insight into all manner of human endeavors. In the language of this book, it is the belief that the knowledge infrastructure works most effectively when use of the infrastructure results in data that can be collected and analyzed to further refine and gain efficiencies in that infrastructure.

If it is not clear by this point, I am not a huge fan of this approach. I've already stated my primary objection on the topic of technology as neutral. What we collect, who we collect it from, and the conclusions we take from data come from socially constructed norms and represent the individual biases of those gathering the data.

For example, data science studies showing gender differences in different topics (for example, sustainable development[15] or software development[16]) are based on a societal definition of gender. In essence, the data does not show a reality but a set of classifications that are developed socially, such as gender identity. If you read that sentence and get a bit angry, convinced that gender is binary—a man and a woman—the anger comes from the fact that what many may have considered set definitions are being challenged.

If you are convinced gender is static and a fact, then perhaps a less political example: race. In the United States, not everyone we consider white today was always considered so. Every 10 years since 1790, the U.S. Census has sought to not only count all Americans, but classify them by race. The Pew Research Foundation provides a fascinating record of the different racial classifications in the census by year.[17] In the 1870 census, many peoples from Asia were simply classified as Chinese. This included people from China, Japan, the Philippines, Korea, India, and Vietnam. In the 2010 census, all of these nationalities had their own entries. It should be noted that before the 1960 census, people were assigned their classification by the census takers. Starting in 2000, people filling out the census could classify themselves as multiple races.

The data from all of these censuses is available and is often used to track things like migration of different peoples from abroad and across the United States. Yet, the definitions of races were changing nearly every census—not through evolution, but through social movements, lobbying, and new understandings. Simply crunching the data will not in and of itself explain these changes. That takes history, that takes anthropology and sociology, that takes reflection and an interrogation not of what can be measured, but what matters.

Take the assertion that by 2040 the United States will no longer be a majority white nation. This fact has been used for everything from calls for greater inclusion in voter rolls, to explaining the rise of white nationalism and the election of Donald Trump, to a rallying call for white supremacists. And yet, the data is not only socially constructed (what is white), but many people assume that what is counted is ethnicity as in skin color, when in fact it is constructed by national origin. The current working definition of white on the census is:

> A person having origins in any of the original peoples of Europe, the Middle East, or North Africa. It includes people who indicate their race as 'White' or report entries such as Irish, German, Italian, Lebanese, Near Easterner, Arab, or Polish.[18]

I'm not sure a white supremacist would agree that someone born and raised in Lebanon or Saudi Arabia would fit their definition of "white."

At this point, a good data scientist would cry foul. Good data science would examine classifiers and account for shifting data definitions. This is absolutely true. If only we were surrounded by just *good* data science—data science that appreciates the entirety of the scientific process. But all too often, decisions are being made from the available data and tools that make accumulating and visualizing data easy. For a book of examples, I refer you back to *Weapons of Math Destruction* (discussed in chapter 3).

Pure data approaches have also led to dubious claims of causality. In his book *Everybody Lies: Big Data, New Data, and What the Internet Can Tell Us about Who We Really Are*,[19] Seth Stephens-Davidowitz shows example after example of insights one can glean by looking through search logs. He diagnoses everything from sexual proclivities to racism by correlating different data sets together—like voting for Donald Trump for president and the number of racist Google searches in key voting groups. These are striking and often compelling correlations, but ultimately narratives: human interpretations of data. Quantitative methods have always been driven by human narratives in the form of hypotheses and even in the choice of which phenomena to study. But now, the fact that human interpretation is involved is either obscured or, worse, unrecognized.

Dataism is more, however, than a set of methodologies. Quantitative and statistical methods to understand everything from the effects of polluted water, to the spread of ideas, to the chances of people dying from cancer have always depended on the collection and interpretation of data—the more the better. Yet these methodologies and corresponding philosophical grounding (positivism, naturalism, constructivism, etc.) have never claimed to be the truth. Let me say that again: science NEVER claims to be the truth. In fact, that is what differentiates science from religion and other belief systems.

At the core of the scientific method and practice is the concept of falsifiability. No matter how much data you collect, no matter how long a theory has stood, one data point that cannot be explained by a scientific "law," and that law is thrown out. The public may believe Darwin presented the truth in his theory of evolution, yet if pressed, he wouldn't have made that claim

All theories such as the theory of evolution (that has been at the heart of modern health care and the development of new antibiotics) to relativity (that is at the heart of the global positioning system you use in your car as well as astronomical explanations of black holes) to germ theory (the very reason for the $1.3 trillion pharmaceutical industry) do not purport to be truth. They claim to be the best explanation of the facts at present. All of these theories, it should also be noted, started out with a creative process whereby scientists imagined possible causes and effects for the phenomena they were studying.

Data science techniques are powerful tools and methods, but they are tools in a larger pursuit of truth and meaning. It is in this larger pursuit, a uniquely human endeavor, that we find understanding and purpose. People are not data processors; they are learners. Just because something can be counted, or something is intended to quantify a certain behavior, does not mean it automatically does so.

As we have seen throughout our winding paths through history from the deck of the CS *Alert*, not only is data a construct, but the entire system of computing and networks that allow for massive mining of data by deep learning algorithms is also socially constructed. Choices we make from the purchase of a television to agreeing to exchange private details for functionality both shape and are shaped by the society we live in. Choices made by politicians, programmers, CEOs, hackers, and citizens—often in uncoordinated ways—have implications for what "truth" emerges from the data trails we leave behind.

Understanding the societal costs of complex tools such as encryption, compression, digital fingerprints, microprocessors, international packet-switched networks, and artificial intelligence can be paralyzing. Yet giving up your agency, your power to make change, or ignoring the consequences of viewing the world as a massive data processor is not an option. We—you and I—have a voice in our future. No technology is inevitable, nor is it neutral. We—you and I—will choose the world to come. But only if we first look unflinchingly into the face of global forces and declare that we are important, and that we will decide.

NOTES

1. James Gleick, *The Information: A History, a Theory, a Flood* (London: Fourth Estate, 2011).

2. Janko Roettgers, "Netflix Subscribers Streamed Record-Breaking 350 Million Hours of Video on Jan. 7," *Variety,* March 8, 2018, https://variety.com/2018/digital/news/netflix-350-million-hours-1202721679.

3. Jack Beckett, "Nazi Germany Inventions the US Made Use of After WWII," War History Online, September 25, 2017, https://www.warhistoryonline.com/world-war-ii/8-nazi-german-inventions-the-us-made-use-of-after-wwii-2.html.

4. "Operation Osoaviakhim," *Wikipedia*, last modified July 11, 2020, https://en.wikipedia.org/w/index.php?title=Operation_Osoaviakhim&oldid=932630274.

5. John Buydos, "Science Reference Guides: Russian Abstract Journals in Science and Technology," Library of Congress, last modified March 1, 2004, https://www.loc.gov/rr/scitech/SciRefGuides/russian2.html.

6. For more information on VINITI, particularly in the health sciences, see Valentina Markusova, "All Russian Institute for Scientific and Technical

Information (VINITI) of the Russian Academy of Sciences," *ACTA Informatica Medica* 20, no. 2 (2012): 113–17, http://doi.org/10.5455/aim.2012.20.113-117.

7. Defense Technical Information Center, accessed July 21, 2020, https://discover.dtic.mil.

8. Jon Marcus, "Most Americans Don't Realize State Funding for Higher Ed Fell by Billions," *PBS News Hour*, February 26, 2019, https://www.pbs.org/newshour/education/most-americans-dont-realize-state-funding-for-higher-ed-fell-by-billions.

9. Office of Public Affairs, "General Fund Budget Snapshot," University of Michigan, accessed July 21, 2020, https://publicaffairs.vpcomm.umich.edu/key-issues/tuition/general-fund-budget-tutorial. For the sake of disclosure, at my own institution, the University of South Carolina, 10.8 percent of our budget in 2018 was from the state, 23.2 percent was from external funding (this includes gifts and alumni donations), and 29.4 from tuition. *University of South Carolina Budget Document: Fiscal Year 2018–2019*, presented to the Board of Trustees, Columbia, South Carolina, July 2018, https://www.sc.edu/about/offices_and_divisions/budget/documents/fy2019botdoclegacy.pdf.

10. Camilo Maldonado, "Price of College Increasing Almost 8 Times Faster Than Wages," *Forbes*, July 24, 2018, https://www.forbes.com/sites/camilomaldonado/2018/07/24/price-of-college-increasing-almost-8-times-faster-than-wages/#7a35aa4566c1.

11. "20 Years of Tuition Growth at National Universities," *U.S. News*, September 20, 2019, http://usanews.over-blog.com/2019/09/20-years-of-tuition-growth-at-national-universities.html.

12. David Brooks, "The Philosophy of Data," *New York Times*, February 4, 2013, https://www.nytimes.com/2013/02/05/opinion/brooks-the-philosophy-of-data.html.

13. Yuval Noah Harari, *Sapiens: A Brief History of Humankind* (New York: Harper Perennial, 2014).

14. Yuval Noah Harari, *Homo Deus: A Brief History of Tomorrow* (New York: HarperCollins, 2018).

15. Claudia Abreu Lopes and Savita Bailur, *Gender Equity and Big Data: Making Gender Data Visible* (Report, UN Women, 2018), https://www.unwomen.org/en/digital-library/publications/2018/1/gender-equality-and-big-data.

16. Hannah Augur, "Big Data Reveals Big Gender Inequality," Dataconomy, March 8, 2016, https://dataconomy.com/2016/03/big-data-reveals-big-gender-inequality.

17. “What Census Calls Us,” Pew Research Center, February 6, 2020, https://www.pewsocialtrends.org/interactives/multiracial-timeline.

18. Daniel Luzer, “Primary Sources: The 1940 Census on ‘White,’” *Mother Jones*, August 19, 2008, https://www.motherjones.com/politics/2008/08/primary-sources-1940-census-white.

19. Seth Stephens-Davidowitz, *Everybody Lies: Big Data, New Data, and What the Internet Can Tell Us about Who We Really Are* (New York: HarperCollins, 2017).

Data Waypoint

The Span of the Knowledge Infrastructure

Before I jump into a discussion of media, let me take a moment and examine the idea (and history) of the knowledge infrastructure in more depth. Doing so will explain why my writing on media will oftentimes not align directly with some professional and academic definitions of media and mass media.

The boundary between data and media is increasingly porous and fuzzy. When my first child was born in 2000, my wife and I pledged to never allow our children to have TVs in their bedrooms. Now that my youngest son is a senior in high school, we have kept our promise—and still failed miserably in our parenting goal. There is no device in his room that could be called a television, but he can watch hundreds of thousands of shows and movies on his phone, his school-supplied laptop, and the desktop computer where he does most of his gaming. His teachers also regularly assign YouTube videos as part of coursework.

This is the reality of the knowledge infrastructure today. It is in flux and traditional concepts, such as the definition of a television or the difference between film and radio and community forum, are dissolving. One of the hot new genres of audiobooks as I write this is the graphic audio book—the adaptation of a graphic novel in audio form. Think on that for a moment.

As we have seen in the first section of this book, this is not a new situation. The knowledge infrastructure is always being created and destroyed.

It is conceptualized, documented, and regulated in radically different ways at different times in history, and indeed in different nations and by different peoples.

Recall that the knowledge infrastructure is comprised of the people, technology, sources, and policies a society has in place to learn and find meaning. It might be easy to mistake this for how facts and news are moved around a population. It is not. The infrastructure includes news and newspapers, but it also includes libraries and books. Fiction and nonfiction. Theater, debate, advertising, religious institutions, and all the ways in which we encounter and transform information into knowledge to move our lives forward. It includes our formal education system in primary, secondary, and post-secondary schools. However, it also includes informal learning through things like YouTube and apprenticeships and social connections like your neighbors and your Facebook friends.

John H. Falk of the Institute for Learning Innovation at Oregon State University knows this well. He demonstrated the limitations of formal education with science proficiency. He points out that when fourth-grade students in the United States are compared with their peers around the world, they rank seventh in scientific understanding. When again tested in eighth grade, they rank 23rd. Yet when science knowledge is evaluated and ranked globally after formal education (after high school, after college), the United States ranks first.[1]

Think on that for a moment. As students proceed in their studies in the United States, science instruction becomes more specialized and more formalized—and performance goes down. In the absence of formal education, it trends higher.

Falk's explanation for what he calls U-shaped performance comes down to learning that is anticipated and regimented versus education driven by personal passion and education that is informal. He attributes the improvement in science knowledge to learning on the job and access to "informal science venues."[2] The biggest of these venues in his work? Libraries.

The knowledge infrastructure is always evolving, changing right along with advancement of civilization. Sometimes in history it was formal and apparent, such as with the advent of scribes in ancient Egypt. In the Egypt of the pharaohs, the knowledge of writing was limited to a special social class—the scribe. Scribes were the only people permitted to read and write, and they documented everything:

> They also employed scribes to record everything from the stocks held in the stores for workers, the proceedings in court, magic spells, wills and other legal contracts, medical procedures, tax records and genealogies. Scribes were central to the functioning of centralised administration, the army and the priesthood and in truth very little happened in ancient Egypt which did not involve a scribe in some manner.[3]

In Egypt, the knowledge infrastructure was an important mechanism of deliberate control.

Over the millennia the visibility and study of the infrastructure has varied from invisible to evident and studied: the rise of the printing press in Europe, for example, called attention to how information was shared, and with whom. We can also see the effects of a knowledge infrastructure in the establishment of universities in the city-states of Renaissance Italy.

With the large influx of classical works from Iberian Spain and the Middle East during the Catholic Crusades, rulers in places like Florence in Tuscany saw opportunities for increasing their wealth and power. Jewish scholars were hired to translate works while monks and clerics, who were literate and not burdened with the tasks of everyday survival, were hired as tutors. Eventually rich town elders combined their investments to bring together the tutors in buildings and furnish them with libraries. These were the first universities.

Much of the materials being captured and translated in the new universities came from Muslim and Indian libraries that had saved ancient Greek texts from the destruction of the Great Library of Alexandria and where discoveries in architecture, philosophy, and mathematics continued. These recovered texts led to new architectural marvels as well as a renewed interest in science and engineering.

Outside of Renaissance Europe and Asian countries like China and what is now modern-day Korea, the knowledge infrastructure was largely invisible and primarily an oral system for passing on stories and skills. In Western Europe during the Dark Ages it was the Catholic Church that took on great responsibility for (some would say used as means of control) the maintenance of the knowledge infrastructure. Monasteries served as archives for texts that were copied and preserved. Priests became scholars to teach.

In China, the knowledge infrastructure was also deliberately managed and groomed. An imperial bureaucracy documented and regulated vast areas of territory. New paper-making technology developed in the Song Dynasty met with the Chinese-developed printing press to produce and widely distribute books and materials.

The knowledge infrastructure became clearer with the development of scholarly publications in France in the mid-1700s, and the focus on the information superhighway (a vision for an expanded internet) in the late 1990s. The knowledge infrastructure was on full display in the Second World War, not only in the computer and Operation Paperclip referenced previously, but the rise of microfiche as a way of preserving cultural items and significant books through a radical new technology, microfilm. There was a time when it was assumed the totality of printed works would be available to everyone. The world in a briefcase—the internet before the internet.

However, the visibility of the knowledge infrastructure goes beyond just technologies to connect people with ideas (and more importantly, to other people). It happens in law. As we will see in the next chapter, governments used taxation and laws of libel and sedition to attempt to regulate the spreading of ideas.

Today the narrative around copyright and patents is one of ownership—who benefits from a piece of intellectual property. However, these laws developed in most nations as a way to increase the circulation of ideas. Authors were given a copyright for a work so that they could gain a profit and then give it away. Patents, too, were way of giving the originator of an idea a limited monopoly on that idea as a reward for making it known, documented, and available to the world. Even trademark policy that covers things like institutional names, brands, and logos originated as a way to prevent confusion in the public, not fence things off from reuse.

We will revisit these ideas in the pages ahead. But for now, take this as an introduction to my discussion of media. Mass media, yes, but also popular media, social media, and the mediums of books and radio and mobile phones.

NOTES

1. John H. Falk, "The Science Learning Ecosystem," paper presented at Public Libraries & STEM: A National Conference on Current Trends and Future Directions, Denver, Colorado, August 2015, https://www.lpi.usra.edu/education/stemlibraryconference/events/Falk_%20Learning_Ecosystems_PRINT.pdf.

2. John H. Falk, "STEM Learning: A Lifelong, Life-Wide View," paper presented at Public Libraries & STEM: A National Conference on Current Trends and Future Directions, Denver, Colorado, August 2015, https://www.lpi.usra.edu/education/stemlibraryconference/presentations/Thursday/Reflection-Discussion/Falk.pdf.

3. Jenny Hill, "Scribes in Ancient Egypt," Ancient Egypt Online, accessed July 21, 2020, https://ancientegyptonline.co.uk/scribe.

Part II

MEDIA

All media exist to invest our lives with artificial perceptions and arbitrary values.

—Marshall McLuhan

I don't necessarily agree with everything I say.

—Marshall McLuhan

7

Propaganda

From Influencing Minds to Manipulating Brains

When the CS *Alert* steamed away from Dover early on August 15, even the captain didn't realize the full nature of his mission. He was to cut the submarine telegraph cables of the Germans, that was clear. What wasn't clear was that while the *Alert* was crippling the communications of the enemy, it was simultaneously increasing the communication power of the British Empire.

Once again, we see the *Alert* and her crew as a nexus of world events. I have talked about her mission's military purpose—driving German communication to channels that could be intercepted for intelligence activities—and how this led to the rise of encryption. However, this military mission was only part of the picture. A declaration of war between the British Empire and Germany also cut a vital trading relationship between both countries. With war, the top trading partner of each country were now enemies.

To make up for the loss of valuable supplies of goods and commerce, both Germany and Great Britain launched concerted efforts to court the United States into favorable trade agreements and, as the war went on, to enter the conflict on their side.

What follows is the story of how the ferocity of the military campaign of World War I was matched by an equally fierce communication battle fought with the twin weapons of mass communications: propaganda and censorship. The lessons of this battle would shape postwar Germany, and the communication reality of the Second World War in a new age of mass media. Inflated stories of German atrocities during the first war led to a slow understanding of the horrors of the Holocaust in the second. After the Second World War, new public and scientific understandings of propaganda and media manipulation would shape the Cold War, our current perceptions of "fake news," and the recent adoption of gamification and the direct manipulation of dopamine cycles in the brain. In essence, this chapter is the story of how the knowledge infrastructure was weaponized.

FROM WELLINGTON TO WASHINGTON

The crew of the *Alert* were not the only people unaware of the full dimensions of their mission or that it had a propaganda dimension. Even in the British government, few members of parliament knew about the launch of a wartime propaganda office, Wellington House, that had been established to take advantage of Britain's dominance of telegraphy and court American sympathy and support through a clandestine propaganda campaign.

Propaganda was certainly not new in 1914. The use of media to manipulate a populace's views and actions is as old as language itself. The term dates back to missionary work of the Catholic Church:

> In 1622 Pope Gregory XV created in Rome the Congregation for the Propagation of the Faith. This was a commission of cardinals charged with spreading the faith and regulating church affairs in heathen lands. A College of Propaganda was set up under Pope Urban VIII to train priests for the missions.[1]

For much of history, propaganda as a term and as an idea was seen as a generally positive thing. It was part of the duty of a government, the thinking went, to bring together citizens in a sense of patriotism, just as it was the duty of the church to bring souls to God. This positive view would be transformed in the Great War to where today, propaganda is seen as uniformly negative—founded not in the construction of a national narrative, but in the manipulation of citizens. The reason for this turnaround can be attributed in large part to German attempts to influence U.S. foreign policy.

The German efforts to influence Americans were not clandestine and openly targeted German-speaking citizens and recent immigrants to the United States. This was not a small group. "Germans were the largest non-English-speaking minority group in the U.S. at the time. The 1910 census counted more than 8 million first- and second-generation German Americans in the population of 92 million."[2] German Americans worshiped in German, drank German beer, and German families living in cities like Cincinnati and Chicago could even send their children to German-speaking public schools.

The wartime German government reached out to this population through posters, speakers, editorials, and pamphlets promoting the message that the United States had a moral obligation to support the fatherland. The German strategy failed so completely that such open campaigns of propaganda would be forever tainted as crass manipulation.

Why did the effort fail? Well, for one thing, few nations appreciate being told what their obligations are by foreign powers. But beyond this, the German government seemed to be doing its best to undermine their own message. Two specific German acts pushed Americans toward the British: the sinking of the *Lusitania*, and the Zimmermann Telegram. I have already discussed the Zimmermann message, the intercepted telegram in which Germany sought an alliance with Mexico in exchange for a promise of the return of Texas, New Mexico, and Arizona.

The RMS *Lusitania* was a civilian British ocean liner. On May 7, 1915, a German U-boat sank the ship 11 miles off the southern coast of Ireland, killing 1,198 passengers and crew. The Germans defended their right to sink the ship because the *Lusitania*, they claimed, was smuggling weapons in a declared conflict zone. This was flatly denied by the British . . . until 1982, when the head of the British Foreign Office admitted that there was a large amount of ammunition in the wreck. However, at the time, the story that took hold in the U.S. popular press was the death of civilians at the hands of a militarized nation.

The final reason why the German propaganda campaign failed was that, simply put, the Brits were better at it. And that brings us back around to the *Alert* and the cutting of cables.

The British plan to influence U.S. policies was founded on the idea that direct messaging and overt recruitment, like the German method, would be unsuccessful. The British planned out a covert campaign. At the heart was the control of telegraph cables which, at the time, meant control of the news: "Thanks to their control of the direct cable communications between Europe and North America, the British also monopolized the news, and news was to be the basis of the British propaganda campaign—all of it carefully censored and selected, of course."[3]

British censors would control the stories wired to the United States. Negative stories were stopped, positive stories let through, and some stories were either amplified or fabricated altogether. Most of this last set of stories centered on German atrocities, supporting a theme of Germans as militarized "Huns." American newspapers, the primary mass media vehicle in the United States at the time, would pick up and report stories as they always did, unaware that the British had, in effect, replaced the wire services with a pro-British feed. With the British fingerprints of manipulation concealed, as far as the American papers were concerned, they weren't spreading propaganda, but simply reporting the news. "It was essential to disguise from the American people the fact that the massive bulk of paper material they were receiving from Britain about the war—pamphlets, leaflets, cartoons, and even the news itself—was emanating from Wellington House under Foreign Office guidance."[4]

By 1917 when the United States entered the war, stories on the *Lusitania*, the Zimmermann Telegram, and the heavy pro-British bias in news, had helped convince then President Woodrow Wilson and his government to enter the war on the side of the British.

FROM DEMONIZING THE HUNS TO DEFINING AMERICA

The United States entered the war well aware of the importance of propaganda. However, the messages of the U.S. government differed from the British approach in two key ways: the government developed propaganda for domestic as well as foreign audiences, and it was not clandestine, but deliberately open:

> A week after declaring war, the Americans set up their own propaganda organization, the Committee on Public Information (CPI), under the direction of George Creel, a journalist and supporter of the president. This body was responsible for censorship and propaganda, although Creel was more interested in "expression rather than suppression." He later described its work as "a plain publicity proposition, a vast enterprise in salesmanship, the world's greatest adventure in advertising." The Creel Committee was divided into two sections, the Domestic, which attempted to mobilize America for war, and the Foreign,

> subdivided into the Foreign Press Bureau, the Wireless and Cables Service, and the Foreign Film Service.[5]

Much of the power we attribute these days to Hollywood's motion picture industry started in the First World War. The European conflict cut distribution of foreign films when Germans and Brits had fewer films to distribute due to rationing and scarcity of materials. This left the U.S. film industry, which by 1914 had relocated from New York to California, with ample room to grow. The emerging studios were only too happy to work with the government to produce films that promoted U.S. engagement in Europe, and the U.S. government needed as much propaganda help as it could get.

In World War I the rising mass media met the new realities of total war. No longer would battles be fought by professional armies in designated areas. World War I saw mass conscription of civilians into national armies. Battles were fought in cities. Stories of the horrors of war now affected workers and politicians alike. Newspapers and films would not only be used to justify these changes, but to place blame and, in the United States, link these changes to a larger national narrative.

American propaganda in World War I still shapes the country today, particularly the decision to use propaganda to create a unifying national narrative:

> Many of the CPI's [Committee on Public Information] staff saw their appointment as an ideal opportunity to promote an ideology of American democracy at a time when America itself was undergoing significant social transformations, such as the growth of cities and the closing of the frontier (which in turn affected immigration). Such an ideology could therefore provide a unifying cohesion for a country as diverse as America at a time of war and social change.[6]

Much of what pundits often attribute to the founding fathers and much of the language around American exceptionalism can be traced back to this decision.

It is important to note that, while the U.S. effort under Creel began with the highest ideals, it was an altogether pragmatic affair. As the need to mobilize American support for the war increased, so did the use of more manipulative techniques. High-minded slogans were quickly augmented with stories of "Hun" atrocities.

What happened after this darker turn shows just how effective, and devastating, propaganda can be. John Barry, in his book *The Great Influenza*, shows just how:

> Socialists, German nationals, and especially the radical unionists in the International Workers of the World got far worse treatment. The *New York Times* declared, "The IWW agitators are in effect, and perhaps in fact, agents of Germany. The Federal authorities should make short work of these treasonable conspirators against the United States." . . . What the government didn't do, vigilantes did. There were the twelve hundred IWW members locked in boxcars in Arizona and left on a siding in the desert. There was IWW member Frank Little, tied to a car and dragged through streets in Butte, Montana, until his kneecaps were scraped off, then hung by the neck from a railroad trestle. There was Robert Prager, born in Germany but who had tried to enlist in the navy, attacked by a crowd outside St. Louis, beaten, stripped, bound in an American flag, and lynched because he uttered a positive word about his country of origin. And, after that mob's leaders were acquitted, there was the juror's shout, "I guess nobody can say we aren't loyal now!" Meanwhile, a *Washington Post* editorial commented, "In spite of excesses such as lynching, it is a healthful and wholesome awakening in the interior of the country," . . . All this was to protect the American way of life.[7]

This barbarianism was in the United States of America, and in the name of patriotism. The propaganda efforts of the United States and Great Britain would have lasting effects beyond the war.

FROM THE AMERICAN DREAM TO THE NAZI NIGHTMARE

At the end of World War I propaganda became associated with the worst of human nature and led to the rise of Adolf Hitler and the Nazis. Postwar investigations in the United States surfaced the secret British propaganda campaigns. Many Americans felt manipulated:

> The conclusion was that the United States had indeed been duped into becoming involved on the Allied side, particularly by secret British propaganda emanating from Wellington House. A series of historical investigations by learned scholars reinforced what was fast becoming a legendary belief in the power of propaganda. The debate was, however, seized upon by isolationist elements in American politics who now argued for non-involvement in European affairs and for Americans to be on their guard against devious foreign propaganda.

> Indeed, such was the degree of American sensitivity to foreign propaganda that in 1938 the Foreign Agents Registration Act was passed by Senate requiring the registration with the US government of all foreign propagandists operating on American soil. The act remains in force to this day.[8]

German generals used the success of American and British propaganda efforts as an excuse for the loss:

> The argument ran as follows: the German armies were not defeated on the field of battle; Germany had not been invaded; indeed Germany had been victorious in the East with the Treaty of Brest-Litovsk (1918). How then did Germany lose the war? Because she was betrayed from within; Allied propaganda had caused a collapse of morale at home; the German armies had therefore been "stabbed in the back." This thesis was, of course, used by right-wing elements in the Weimar Germany of the 1920s to "prove" a Jewish-Bolshevik conspiracy that was to help Hitler to power in 1933.[9]

Hitler himself studied the propaganda techniques and campaigns of the Great War. They became his playbook in his rise to power. In the years leading to the Second World War, Hitler used propaganda to demonize Jews and communists while simultaneously inventing the mythology of the Aryans and the Third Reich.

Under Hitler and the rising fascist wave in Europe post–World War I, the knowledge infrastructure once again came into focus, infusing the whole system with the twin tools of media manipulation: propaganda and censorship. Books were written to promote Nazism, while others were burned to limit the ideas that citizens had access to. Just as the United States had taken the opportunity of the war to create a national narrative in a time of change, Hitler oversaw the construction of a new narrative in Germany.

The idea that Germany lost the First World War (and the resulting economic collapse) because of traitors in the population was put forth in one of the most effective pieces of propaganda in history, the book *Mein Kampf*. Nazi propaganda simultaneously built up the concept of the master race and dehumanized Jews, communists, Gypsies, homosexuals, and immigrants. Lists of acceptable books were circulated to book sellers, and materials by prominent Jewish authors were confiscated. Mass rallies were not so much organized populace events as staged dramas for filming.[10]

Hitler and the Nazis didn't just restrict their propaganda efforts to Germany. In the 1930s, Nazi Germany ran extensive campaigns in the United States to garner sympathy for the German cause. Like in World War I, the Nazis used German-speaking social clubs, also known as Bunds. Unlike in World War I, however, the campaigns were better received by an isolationist American public. The goal of the campaigns, like the America First campaign of the 1930s, was less about inviting the United States to join the Axis powers when war came, and more about trying to keep the United States out of the war altogether.[11]

In Italy, Mussolini built strong propaganda campaigns around dissatisfaction with the results of the First World War.

> The Italians in particular were furious. They had entered the war on the Allied side in 1915 under the secret Treaty of London in return for post-war territorial gains in south-eastern Europe that were now being denied them by the principle of self-determination. They left Paris disappointed and disillusioned, seized Fiume (Trieste) in a clash with newly-created Yugoslavia and, in a wave of nationalist euphoria, began the swing to the right that saw Mussolini appointed Prime Minister in 1922. Wartime propaganda had played a significant part in Mussolini's rise and he himself was to convert the lessons of the wartime experience into peacetime use.[12]

Mussolini didn't just focus on mass media, but many parts of the knowledge infrastructure. Once Hitler and Mussolini came to power, they quickly took hold of their respective nations' education systems. School became as much about a new narrative of nationhood as reading and writing.[13] I once had a librarian in the Italian Ministry of Education tell me that the last time they had strong school libraries in Italy was under the fascists.

Mussolini established a High Commission for the press in the spring of 1929. The commission insisted that it promoted a free and open press, though it did reserve the right to an exception for "any activity contrary to the national interest."[14] In 1935 Mussolini's son-in-law, Galeazzo Ciano, formed the Undersecretariat for the Press and Propaganda, later renamed the Ministry of Popular Culture. It became a template later adopted by Joseph Goebbels in the Nazi Party as Germany and Italy formed ever-closer bonds and greater collaboration on the road to war—a war that started on September 1, 1939, when Nazi Germany invaded Poland.

FROM THE PEOPLE'S WAR TO AUSCHWITZ

Two days after Germans poured into Poland, on September 3 the United Kingdom and France declared war on Germany. Once again, the British ramped up their propaganda and censorship machine. This time, however, rather than focusing on the enemy and atrocities, the country focused on its own citizens. The message and public narrative were that the British people were about to fight "the People's War."[15] Unlike World War I, which could be seen as battling empires (the British Empire, Austro-Hungarian Empire, and Ottoman Empire), World War II would be a war waged by and for all of the citizens of the Axis and the Allied countries:

> Before the war, the working man and woman had been largely caricature figures of fun. The People's War, however, demanded that they were now taken seriously and in many respects the strict censorship of the pre-war years, as exercised by the British Board of Film Censors (BBFC), was now relaxed in its treatment of social issues.[16]

Also, unlike World War I, the British had no monopoly over the media as they did with the telegraph. The knowledge infrastructure in the late 1930s and 1940s was no longer dominated by wire services and the written word. This would be a war fought in the papers, certainly, but also on the now-dominant broadcast media of radio and film. This meant that, unlike in the time of telegraphy when Great Britain sat at the center of a global web of communication (and persuasion), this time around that center would be firmly occupied by the United States.

When the United States did enter the war after the bombing of Pearl Harbor in December 1941, a new domestic propaganda operation was also mobilized. Films in World War I were more experiment than mainstay, and radio was limited to wireless telegraphs. In the 1940s, the radio and film industries were massive, and connecting to a patriotic message not only served the country, but served the pocketbooks of these industries as well.[17]

In Nazi Germany, Hitler and Goebbels were fixated on film, requiring all theater-goers to view weekly newsreels about the glory of the Third Reich. Nazis also ruled radio, creating the Reich Radio Society to control a network of 26 stations and serve as "the first and most influential intermediary between movement and nation, between idea and man."

In the United States, President Roosevelt, well versed on the power of mass media from his Depression-era Fireside Chats, let loose the creative potential of the film industry:

> Once the Office of War Information had been set up in June 1942, the US government issued a manual to Hollywood listing the kind of themes that would serve the national effort. Classified as "an essential war industry," Roosevelt stated: "I want no restrictions placed thereon which will impair the usefulness of the film other than those very necessary restrictions which the dictates of safety make imperative." Five themes were identified as needing priority: (1) to explain why the Americans were fighting; (2) to portray the United Nations and their peoples; (3) to encourage work and production; (4) to boost morale on the home front; (5) to depict the heroics of the armed forces.[18]

Beyond film and radio, the Allies developed unique technologies to wage the propaganda war, such as the Monroe bomb. Deployed from a bomber, the Monroe bomb dropped to 1,000 feet then showered up to 80,000 leaflets and newspapers into enemy territory.

I cannot leave a discussion about media and World War II without talking about perhaps the most important message that the mass media carried at the end of the war. As the Allies advanced into German-held territory, they encountered the true horror of the Nazi regime firsthand. They began liberating concentration camps and found "hundreds of thousands of starving and sick prisoners locked in with thousands of dead bodies. They encountered evidence of gas chambers and high-volume crematoriums, as well as thousands of mass graves, documentation of awful medical experimentation, and much more."[19]

The United States Holocaust Memorial Museum estimates that Nazi Germany killed 6 million Jews, around 7 million Soviet citizens, 3 million Soviet prisoners of war, 1.8 million non-Jewish Polish citizens, 312,000 Serb citizens, 250,000 people with disabilities, 250,000 Roma (Gypsies), 1,900 Jehovah's Witnesses, 70,000 repeat criminal offenders, and hundreds or possibly thousands of homosexuals.[20]

In many narratives that emerged in the media at the time, it was only in the fall of Germany and the invasion of Allies that the Nazis' level of evil was discovered. However, the Jewish genocide and mass killings were known to both the Allies and newspapers during the war. In another example of how the knowledge infrastructure is not an unbiased conveyor of facts, but rather a social construct, the death of millions of Jews received little attention during the war itself.

From a combination of anti-Semitism in government, Allied nations' reluctance to raise immigration quotas for rescued Jewish refugees,[21] journalists' reluctance to carry atrocity stories from Germany because of worries of anti-Semitism among their readers,[22] and an attempt to avoid the false information used in World War I,[23] the horrors were slow to spread. It wasn't until irrefutable photographic evidence was released that the world began to understand the sheer horror of the Nazis' "Final Solution."

Unfortunately, the story of the German Holocaust is evidence of another feature of the knowledge infrastructure: facts and narratives are not guaranteed permanence. A 2018 study commissioned by the Conference on Jewish Material Claims Against Germany found that

> nearly one-third of all Americans and more than 4-in-10 Millennials (41 percent) believe that substantially less than 6 million Jews were killed (2 million or fewer) during the Holocaust. While there were over 40,000 concentration camps and ghettos in Europe during the Holocaust, almost half of Americans (45 percent) cannot name a single one—and this percentage is even higher amongst Millennials.[24]

And in places like the Middle East, where countries such as Iran actively campaign to deny the Holocaust ever happened? Only 8 percent of respondents reported that they had heard of the genocide and believed descriptions of it were accurate.[25]

FROM THE COLD WAR TO SOCIAL MEDIA

After World War I, explicit government propaganda operations were suspended. After World War II they were initially stopped, but were quickly revitalized in the face of the Cold War. The United States and Soviet Union both created propaganda and information/misinformation campaigns as the competing superpowers sought global influence.

Yuval Harari, in his book *Sapiens*, outlines three global narratives in the twentieth century: fascism, communism, and liberal democracy. Fascism was effectively discredited with the Second World War. This left communism and liberal democracies as two competing ideologies through much of the second half of the twentieth century.

These narratives had a massive effect on the world order from the 1950s through the fall of the Soviet Union in 1991. They were used as justifications for the Korean and Vietnam Wars. The narratives were transformed into policies and laws like the one that differentiated the spread of scientific and technical information outlined in chapter 6. The Stasi, East Germany's internal intelligence operation, sought to control every aspect of the knowledge infrastructure from education to media to the interaction between family members.

Television as a mass medium joined radio and film, with all the familiar predictions of the death of the former media and utopian promises to educate and bring us all closer together. With the rise of television came the continued evolution of marketing, public relations, and advertising.

Government propaganda was met with the rising sciences of psychology and sociology. Propaganda evolved into concepts of psychological warfare in the early atomic era. Soviet and American alike sought out new means of manipulating citizens—their opponent's and their own.

In 1980 the Cable News Network (CNN) began 24-hour news broadcasts. The advertising business model had been transferred from newspaper newsrooms to TV production suites, and was now headed for cable. News, long a mainstay of television station profits, now had to fill hours of airwaves with news compelling enough to attract viewers and generate advertising revenue. The practice of journalism met the market research of advertising and the science of psychology and communication.

New cable services began to break down mass media markets to viewerships. Three networks fractured into a hundred channels, and the daily menu of programming was deconstructed into targeted specialties. Mornings of children's programs, afternoons of soap operas, the nightly news, evenings of game shows, and a prime time with scripted comedies and dramas became the Comedy Channel, the Cartoon Network, CNN, and MTV. Programming that could capture large national audiences became fewer and far between, driving up the prices of Super Bowl advertisements and live sports—the few programs that still seemed to garner a large real-time audience for advertisers.

No longer could marketers or propagandists depend on select channels to convey their messages. It wasn't enough to manipulate the newspapers or the nightly news. Now the fight between communism and liberal democracy would be played out in music videos and stand-up comedy specials.

Into this fractured mediascape came the internet. And just as Great Britain realized the intelligence potential of the telegraphy network, state actors and non-state actors realized the death of mass media could mean the birth of new forms of influence and propaganda.

In the asymmetrical wars that followed Desert Storm in Kuwait, social media played a significant role in information warfare. The massive cost and bureaucracy necessary in the days of mass media to reach a broad audience could now be accomplished with minimal investment through Facebook, Twitter, and YouTube. What's more, these new social media tools could go far beyond simple broadcasts. Propaganda messages could be followed up by interaction. Discussion boards, comment chains, and private chat groups could all be used to directly interact with microtargeted populations. This approach is particularly powerful because it matches how people learn. People understand their world by seeking out messages they agree with and then engaging in conversations (mostly with themselves as critical thinking). People seek out systems where they have control.

Unlike the overt German propaganda of World War I, internet-based propaganda could now tailor messages. Now propagandists could pick the places they wanted to recruit, the demographics, even know the political preferences of potential sympathizers to their messages. Political parties and terrorist cells alike could tap into the growing data market (legitimate or not) to seek out new supporters. There was little concern about disguising messages because they would be lost in the increasing noise of the public square. Instead of one ad for a political candidate, campaigns could create hundreds of variants market-tested to micro-audiences. Stay-at-home moms in the Midwest with a family income of under $50,000 would hear about the importance of jobs, while white males in New York with a household income of over $150,000 would get the candidate's view on tax breaks. Rather than seeking out a national narrative, campaigns would be playing on differences that would lead to political and economic success.

Social media also has great advantage over mass media in the world of influence: your brain.

FROM DOPAMINE TO GAMIFICATION

Dopamine in the brain is a neurotransmitter. Neurons release dopamine to signal anticipation and prioritize actions. It is, in essence, a chemical that propels us to seek out a given stimulus for an anticipated reward. Dopamine is an essential chemical in learning, and is released to motivate a person to prioritize attention. It does so in anticipation of a reward. If the reward is greater than anticipated, dopamine and other neurotransmitters build a relationship between the stimulus and the reward. If the reward is less, or nonexistent, the connection between the behavior and the reward is decreased or eliminated.

Dopamine and its chemical cousins drive a toddler to walk, a reader to turn the page, and a crack addict to seek out crystalized cocaine. You see an environmental cue (that Krispy Kreme warm donut sign) and dopamine starts pumping, pushing you to get donuts because you have learned that warm donuts are good.

It turns out that there are ways to manipulate and control the dopamine cycle. The easiest way to trigger dopamine release is through drugs. Morphine, alcohol, nicotine—a whole host of drugs will overstimulate dopamine release. This is part of the cycle of addiction. Addicts don't seek out their substance of abuse because they like it, but because their brain is fixated on finding it.

Another means is through novelty and seeking out a stimulus that rewards a behavior. When you win at a game, you get a boost of dopamine, and therefore find the game pleasurable, and want to play again. It is important, however, that for increased and sustained engagement, randomness is involved. If you do the same actions for the same reward, it becomes a habit. This is why we jump to check our mail or texts when we hear a ping, but that doesn't lead to us feeling a high—we are conditioned, but not rewarded. However, in stimuli like games, the outcomes are varied. You play a game all the time not because you always win, but because you know you may not, so the reward is even "sweeter."

This is the chemical foundation of a concept called gamification. To "gamify" a task is to associate novel rewards and achievements into that task, which activates this dopamine cycle. A classic example is something many readers will already have on their wrists. For some it may be an Apple Watch, for others a Fitbit, but increasingly people are engaged in daily monitoring of their physical exercise.

The benefits of exercise are nothing new, and are widely known. Yet many of us don't do enough. Enter the activity monitor, mostly in the form of step counters. It is one thing to know you should exercise; it is quite another to know just how close you are to a daily goal of 5,000 steps, or an achievement (closing your rings), or how many more steps you have taken than a friend.

FROM FITBITS TO ELECTION INTERFERENCE

Gamification, like all the technologies and business models discussed here, is often presented as a positive development. High school and college classes have been gamified to drive greater engagement of students. Health applications like the Fitbit drive healthier living. Everything from saving money for retirement, to quitting smoking, to spending less time on devices have been gamified to promote positive behaviors. But, as with the other technologies and business models, the important question isn't what something does, or how it does it, but who is choosing the behaviors to influence. This human drive (for profit or progress) get baked into the tools.

It also turns out that social media plays into this concept. Making a post may not trigger a dopamine response, but getting a like does (or a heart on Instagram, or a retweet on Twitter). Getting a lot of likes increases the reward. Soon, such activities can influence behavior. "This type of posts got more likes last time, let me do more of those." Look across your phone—how many apps are rewarding you with likes or hearts, or coins, or lives?

> In an unprecedented attack of candor, Sean Parker, the 38-year-old founding president of Facebook, recently admitted that the social network was founded not to unite us, but to distract us. "The thought process was: How do we consume as much of your time and conscious attention as possible?" he said at an event in Philadelphia in November. To achieve this goal, Facebook's architects exploited a "vulnerability in human psychology," explained Parker, who resigned from the company in 2005. Whenever someone likes or comments on a post or photograph, he said, "we . . . give you a little dopamine hit." Facebook is an empire of empires, then, built upon a molecule.[26]

While the dopamine cycle is not equivalent to brainwashing or addiction, it can change behavior over time. Every day as I drive to work, I plug my office address into Waze, a GPS app. Why every day? Because there are multiple routes, and traffic conditions can change day to day. So, I have Waze find me the best route. The first few times Waze saved me time; I felt a positive reward. Now using the app is not only a habit, but one might honestly ask if I am using the app to find my way, or if the app is using me to serve up ads? If I use other features of the app, such as reporting traffic or the position of a police car or add pictures or reviews, I get points. I get a weekly report that tells me how many people I helped. Click me, use me, pay attention to me.

Gamification and dopamine cycles are now part of a whole host of attention-getting and manipulation techniques. Tie them into longstanding propaganda techniques,[27] like "bandwagoning," or peer pressure, and the effectiveness of manipulation is increased. Fox News and MSNBC actively extol that they are part of a nation or a group building peer pressure on TV and online. With "unreliable testimonials," famous faces are used to disseminate an idea. Fuse bandwagons and these testimonials with social media, and you get the modern-day influencer.

Nations are also seizing on these new aspects of digital media to weaponize the knowledge infrastructure. Russia has deployed chat bots to influence national elections in the United States, the United Kingdom, and Germany since at least 2016. Chat bots are automated pieces of code that act as users of social media. They scan millions of feeds on Facebook and Twitter looking for key phrases, and then respond to users with a message intended to influence. Hackers imitate friends you know to influence behavior or harvest your identity for resale. Propaganda has gone from sending out a message or dominating the media to actively manipulating the way you think, based on psychology and neuroscience.

As I move forward with both grounding, and then advocating for a new knowledge infrastructure, knowledge of these supercharged methods of manipulation must be accounted for. A very serious issue in the infrastructure, however, is being able to learn lessons and preserve the evidence of manipulation for the future. And that takes us into the role of memory organizations in media.

NOTES

1. Ralph D. Casey, *EM 2: What Is Propaganda?* (1944; pamphlet reproduced by the American Historical Association), https://www.historians.org/about-aha-and-membership/aha-history-and-archives/gi-roundtable-series/pamphlets/em-2-what-is-propaganda-(1944)/the-story-of-propaganda.

2. Robert Siegel and Art Silverman, "During World War I, U.S. Government Propaganda Erased German Culture," *NPR*, April 7, 2017, https://www.npr.org/2017/04/07/523044253/during-world-war-i-u-s-government-propaganda-erased-german-culture.

3. Phillip M. Taylor, *Munitions of the Mind: A History of Propaganda from the Ancient World to the Present Day* (Manchester, UK: Manchester University Press, 2003), 178.

4. Ibid., 177.

5. Ibid., 183.

6. Ibid., 184.

7. John M. Barry, *The Great Influenza: The Story of the Deadliest Pandemic in History* (New York: Penguin Publishing Group, 2004), 206–7.

8. Taylor, *Munitions of the Mind*, 196.

9. Ibid., 188.

10. Robert M. Citino, "Mussolini and Hitler, Propaganda Partners in Crime," History Net, last modified April 2019, https://www.historynet.com/mussolini-and-hitler-propaganda-partners-in-crime.htm.

11. Lily Rothman, "More Americans Supported Hitler Than You May Think; Here's Why One Expert Thinks That History Isn't Better Known," *Time*, October 4, 2018, https://time.com/5414055/american-nazi-sympathy-book.

12. Taylor, *Munitions of the Mind*, 195.

13. Richard J. Wolff, "'Fascistizing' Italian Youth: The Limits of Mussolini's Educational System," *History of Education* 13, no. 4 (1984): 287–98, https://doi.org/10.1080/0046760840130403.

14. David S. D'Amato, "Mussolini Attempted to Remake the Italian Mind, Taking a Personal Interest in Applying the Twin Tools of Censorship and Propaganda," Libertarianism.org, January 28, 2016, https://www.libertarianism.org/columns/mussolini-press.

15. Jeremy Cronig, "The People's War: England's Wartime Narrative," The Ohio State University History of WWII Study Program, May 18, 2018, https://u.osu.edu/wwiihistorytour/2018/05/18/the-peoples-war-englands-wartime-narrative.

16. Taylor, *Munitions of the Mind*, 218.

17. Television was just beginning, and the war, if anything, stunted the development of the medium, though it did make some contributions. See James A. Von Schilling, "Television During World War II," *American Journalism* 12, no. 3 (1995): 290–303, https://doi.org/10.1080/08821127.1995.10731744.

18. Taylor, *Munitions of the Mind*, 230.

19. Alan Taylor, "World War II: The Holocaust," *Atlantic*, October 16, 2011, https://www.theatlantic.com/photo/2011/10/world-war-ii-the-holocaust/100170.

20. "Documenting Numbers of Victims of the Holocaust and Nazi Persecution," United States Holocaust Memorial Museum, last modified February 4, 2019, https://encyclopedia.ushmm.org/content/en/article/documenting-numbers-of-victims-of-the-holocaust-and-nazi-persecution.

21. Andrew Buncombe, "Allied Forces Knew About Holocaust Two Years Before Discovery of Concentration Camps, Secret Documents Reveal," *Independent*, April 18, 2017, https://www.independent.co.uk/news/world/world-history/holocaust-allied-forces-knew-before-concentration-camp-discovery-us-uk-soviets-secret-documents-a7688036.html.

22. Laurel Leff, *Buried by the Times: The Holocaust and America's Most Important Newspaper* (Cambridge: Cambridge University Press, 2006).

23. "American Response to the Holocaust," History.com, last modified August 21, 2018, https://www.history.com/topics/world-war-ii/american-response-to-the-holocaust.

24. "New Survey by Claims Conference Finds Significant Lack of Holocaust Knowledge in the United States," Claims Conference, accessed July 22, 2020. http://www.claimscon.org/study.

25. Emma Green, "The World is Full of Holocaust Deniers," *Atlantic*, May 14, 2014, https://www.theatlantic.com/international/archive/2014/05/the-world-is-full-of-holocaust-deniers/370870.

26. Simon Parkin, "Has Dopamine Got Us Hooked on Tech?" *Guardian*, March 4, 2018, https://www.theguardian.com/technology/2018/mar/04/has-dopamine-got-us-hooked-on-tech-facebook-apps-addiction.

27. A good list and definition here: "Public Relations and Propaganda Techniques," George Washington University, accessed July 22, 2020, https://gspm.online.gwu.edu/blog/public-relations-and-propaganda-techniques.

8

Memory Organizations

From Weaponized Librarianship to a Digital Dark Age

The United States entered World War I on April 6, 1917. Initially, 107,641 U.S. soldiers were deployed to the European conflict, but that number would grow to 2 million by October 1918 and the end of the war through implementing the draft.[1] In 1917 the American Library Association (ALA) organized a Library War Service to provide books, magazines, and indeed whole libraries to the U.S. war effort; "a second book drive in early 1918 generated 3 million books, many going overseas, others ending up on the shelves of 36 training-camp libraries erected through Carnegie Corporation funding and managed by ALA volunteers across the country."[2]

Up to this point, my examination of media has focused on broadcast and mass media. However, the knowledge infrastructure encompasses a much broader array of sources used to build knowledge. Moreover, the utility of broadcast and mass media persists past the point of transmission or delivery (including speeches and lectures). In this chapter I will look at how the role of print media and even physical objects in the war efforts of 100 years ago to today have changed the very concept of permanence in the knowledge infrastructure.

Why permanence? A vital part of what we know is what we remember. In individuals, this memory is at best imperfect. Metaphors of memory that liken it to how we "store" our past are woefully out of date. Psychologists and neuroscientists talk about memory more as an act of re-creation, where events are not simply summoned from the brain's cold storage, but are instead reconstructed again and again, always influenced by the present, and indeed events that have been experienced since the memories were formed.

Societies, too, have a memory that is shaped into history and constantly recast through the lens of the present. Founding Fathers transform from archetypical wise men to flawed slave owners seeking an imperfect compromise. Christopher Columbus goes from discoverer of America to the first wave of colonial oppression and indigenous genocide. Cultural memory goes in other directions as well. Flawed celebrities become archetypes, washing away impolitic views of women or race or class. Charles Lindbergh is remembered as the brave man who flew across the Atlantic, not so much as the brave man who flew across the Atlantic and supported the Nazi cause before World War II.

There is no clearer evidence of this than the events related to the Black Lives Matter racial protests in the summer of 2020. In the United States, statues of Confederate soldiers and Christopher Columbus were torn down by protestors seeking to remove the literal idolization of oppressors. In Europe, statues of slave traders in the United Kingdom, and of colonizers in Belgium,[3] were also torn down by protestors seeking to end sanitized historical narratives. These transformations in the societal narrative are never unanimous, and therefore never uncontested. The removal of the Confederate flag from the Mississippi state flag and from NASCAR events was met with both supportive declarations and cries of erasing of Southern heritage.

Through all of these transformations, the media plays a role in the weaving and unraveling of narratives. This includes so-called memory institutions that either claim, or are given, a societal mandate to preserve the narratives, decisions, and events of the past. This is a role that puts the people who build and maintain these organizations—libraries, archives, museums, universities, schools—in the unenviable position of either relegated to an often-underappreciated task or literally fighting for their lives as those who seek a new narrative forcibly try to rewrite history.

FROM COLLECTING BOOKS TO COLLECTING INTELLIGENCE

During World War I, the ALA's War Service Committee was focused on providing soldiers with books for pleasure as well as current magazines and papers. Books were largely distributed to training camps for men drafted into the war. The service "distributed approximately 10,000,000 books and magazines; and provided library collections to 5,000 locations."[4] The effort formed the basis of military libraries across the world that serve as public libraries on military bases for soldiers and their families to this day. After the war, the mass of donated books formed the core collection for the American Library in Paris, the largest English-language lending library on the European continent today.[5]

In the lead up to the Second World War, however, American libraries would take a very different role. In 1939, Archibald MacLeish was confirmed as the new Librarian of Congress. He was not, as the ALA pointed out in its opposition to his nomination, a trained librarian, but rather a poet. Yet he quickly took on the mantle of the country's top librarian, and began a campaign against fascism, for U.S. entry into World War II, and for librarians to play an active role in the war effort:

> To MacLeish, libraries were active weapons in the cultural side of the battle; they were propaganda guns of truth—"white" propaganda as opposed to the negative of "black" propaganda epitomized by George Creel in World War I. MacLeish saw librarians as soldiers on the battle-front for ideology who must awaken to their historic tradition of defending mankind against the forces of ignorance.[6]

MacLeish was not the only one preparing librarians for war. On December 12, 1941, just five days after the Japanese bombed Pearl Harbor and one day after Germany and Italy declared war on the United States, the ALA issued a declaration outlining how libraries should respond to the attacks:

1. Officially or unofficially, every library must become a war information center in which are currently available the latest facts, reports, and instructions for public use.
2. The library must supply technical information to industrial defense workers and students.
3. The library must disseminate authentic information and sound teachings in the fields of economics, government, history, and international relations.
4. The library must make available valid interpretations of current facts and events.
5. The library must help to relieve the strain of war by maintaining its supply of recreational reading for men and women, and especially for children.
6. The library must help and support postwar planning.[7]

This was far too modest a set of actions for MacLeish. He sought to weaponize librarianship, and did so at the Library of Congress.[8] He created the Division for the Study of Wartime Communications that monitored foreign propaganda efforts. He supercharged the library's reference services, where librarians provided round-the-clock support for defense agencies, including use of the library's extensive map collections to help plan military operations. Even the library's Music Division produced language-training records for the armed forces.

Perhaps MacLeish's biggest contribution was the creation of the Division of Special Information. The division hired librarians, academics, archivists, and book collectors to pioneer the concept of open source intelligence. These "book people" were trained by the British intelligence services and deployed throughout Europe and China. They collected, digested, and microfilmed foreign newspapers, propaganda, and books. They fed this information to British intelligence and to the newly formed Office of Strategic Services (OSS) that I introduced in chapter 6. OSS, the precursor to today's Central Intelligence Agency (CIA) in the United States, worked closely with the Library of Congress to direct clandestine operations, including sabotage campaigns behind enemy lines.

This work led directly to the creation of what we call information science today. New ways were developed to rapidly describe, classify, and disseminate information. Key facts and ideas were extracted from their physical forms (books, newspapers) and stored for rapid retrieval—solidifying the concept and methods around information as separate from documents.

Masses of print media were acquired, stolen, and saved during and at the end of the war, including a large number of valuable books that the Nazis stole from Jews. The work of the Library of Congress and the OSS was increasingly driven by the idea that the culture of the west must be preserved in times of war. Classics of European literature and scholarship were duplicated or simply taken and deposited by the Library of Congress and academic libraries throughout the United States. By the end of the war, many university libraries and private book collections across Europe were either destroyed or emptied into American libraries, now emboldened with a mission to preserve the cultural foundations of man—a decidedly Eurocentric mission.

Here, once again, we see the knowledge infrastructure being shaped by the immediacies of warfare. As with U.S. development of the scientific and technical information system described in chapter 6, the United States was also taking on a larger and larger role as a provider of information and developer of scientific knowledge. Whereas in World War I most important scientific journals were published in German, now they were being published in English. Scientists like Albert Einstein were driven from Germany to the United States before the Second World War, or were relocated by the U.S. Army afterward. Just as the film and radio industry became part of the wartime propaganda, American libraries, museums, and universities were mobilized to support military ends. This includes the famous "monuments men," who filled American museums with treasures recovered from the war as part of the Monuments, Fine Arts, and Archives program established in 1943 under the Allied forces.

FROM PRINT TO PLACE

The Second World War shocked the arts and cultural world with images of book burnings, looting, and mass bombings of cities. The same impulses that drove MacLeish to try and save the cultural record of Europe now occupied the entirety of the cultural community. Academics, historians, museum curators, librarians, archivists, artists, architects, and scholars of the humanities were convinced of the need to save the world's cultural heritage.

Cultural heritage was not a new concept or phrase developed after the Second World War. It was, however, given a greater sense of urgency. The phrase itself has been succinctly defined as the gift one generation passes to another. Or even more eloquently phrased in *Human Capital*, a book put together to celebrate Italy's hosting of the G7 summit in 2017:

> The word "culture" derives from the Latin predicate *colere*, which refers directly to the art of cultivating and, in a wider sense, to the care of people and communities put into cultivating themselves, their talents, their dreams.
>
> The constant focus on the earth, and the rooting of one's identify in it, are an expression of the sacred relationship that mankind has established, since time immemorial, with Nature. The landscape is transformed into a sort of cultic map against which man measures himself, a mystical projection of his own self through which he generates his representation and his vision of things, with a view to evoking the marvelous works of the universe—an experience analogous to that desire for perfection that one feels when contemplating Mother Earth.[9]

This cultivation is expressed in works of art, literature, drama, music, and architecture. That "cultic map" consists of everything from the *Mona Lisa* to the Beatles' *White Album*. It also includes of the teachings of Plato and significant buildings and locations like the Eiffel Tower, the Taj Mahal, the rock paintings of the Sierra de San Francisco in Mexico, and over 1,100 historic pieces of architecture and important natural sites such as the Kenya Lake System in Africa's Great Rift Valley.[10]

Postwar, the world sought to avoid the mass destruction of cities like Dresden and Hiroshima and put in place many international structures to promote cooperation and diplomacy over open conflict. In November 1945, 37 countries founded the United Nations Educational, Scientific and Cultural Organization (UNESCO) as part of the newly founded United Nations. It was created on the belief that "the new organization must establish the 'intellectual and moral solidarity of mankind' and, in so doing, prevent the outbreak of another world war."[11] UNESCO's mission was to promote education, science, and culture—in essence, peace through improving key aspects of the knowledge infrastructure.

In 1972, UNESCO adopted the Convention Concerning the Protection of the World Cultural and Natural Heritage. It defined cultural heritage as:

> **monuments**: architectural works, works of monumental sculpture and painting, elements or structures of an archaeological nature, inscriptions, cave dwellings and combinations of features, which are of outstanding universal value from the point of view of history, art or science;
>
> **groups of buildings**: groups of separate or connected buildings which, because of their architecture, their homogeneity or their place in the landscape, are of outstanding universal value from the point of view of history, art or science;
>
> **sites**: works of man or the combined works of nature and man, and areas including archaeological sites which are of outstanding universal value from the historical, aesthetic, ethnological or anthropological point of view.[12]

The convention also defined natural heritage as:

> **natural features** consisting of physical and biological formations or groups of such formations, which are of outstanding universal value from the aesthetic or scientific point of view;

> **geological and physiographical formations** and precisely delineated areas which constitute the habitat of threatened species of animals and plants of outstanding universal value from the point of view of science or conservation;
>
> **natural sites** or precisely delineated natural areas of outstanding universal value from the point of view of science, conservation or natural beauty.[13]

The 1972 convention had significant goals that affect our look into the knowledge infrastructure. The first is that it expanded the concept of cultural heritage from the art of manmade objects to natural features and places of cultural significance. It also obligated the signatory states to the preservation of these sites, and it sought to limit the impact of warfare on these places: "Each State Party to this Convention undertakes not to take any deliberate measures which might damage directly or indirectly the cultural and natural heritage . . . situated on the territory of other States Parties to this Convention."[14]

This convention has influenced military policy since 1972, including when in 2019 U.S. President Donald Trump threatened to bomb over 52 Iranian targets, including significant cultural sites:[15]

> "They're allowed to kill our people. They're allowed to torture and maim our people. They're allowed to use roadside bombs and blow up our people," Trump told reporters aboard Air Force One on his way back to Washington, D.C, from his Mar-a-Lago resort in Palm Beach, Florida. "And we're not allowed to touch their cultural site? It doesn't work that way."[16]

Trump backed away from his threat after significant outcry from legislators and the international community.

The work of UNESCO is also an excellent example of how the knowledge infrastructure is not neutral but represents the biases and agendas of the people building it. In 1994 UNESCO had to revise the 1972 conventions—not in definitions or obligations, but in perspective:

> Twenty-two years after the adoption of the 1972 Convention Concerning the Protection of the World Cultural and Natural Heritage, the World Heritage List lacked balance in the type of inscribed properties and in the geographical areas of the world that were represented. Among the 410 properties, 304 were

> cultural sites and only 90 were natural and 16 mixed, while the vast majority is located in developed regions of the world, notably in Europe.[17]

Today the convention is far more representational of non-European sites.

FROM MICROFILM TO MICROPROCESSORS

How did the list of cultural heritage sites become so skewed toward Europe in the first place? The simple answer is a clear bias in the west toward, well, the west. MacLeish wanted the Library of Congress involved in World War II because he shared a belief that all culture and society is based on western thought. For many in the United States and Europe, the march to civilizations goes from Greece to Rome to the Renaissance to the Enlightenment—all western advances that either ignored the rich history and contributions of Asia and eastern philosophy, or worse, were directly antagonistic toward it (read anti-Semitic and racist).

This belief that the United States is established on a European foundation, often phrased as Judeo-Christian (with all due irony given the Middle Eastern origins of both Judaism and Christianity), was not the only reason that work in cultural heritage tends to focus on Europe. In the United States, for example, while many museums and archives would claim a strong cultural heritage tradition, most libraries and many science-oriented museums (natural history, science centers), would not. Librarians were, and are, trained as information providers, with the strong influence of the social sciences. This was also evident in how the United States sought to preserve the world's cultural heritage during and after the Second World War. Whereas the Europeans talked about collection and preservation of artifacts, the United States pushed technology as the solution.

The primary technology of the day was microfilm. Microfilm is the use of highly detailed photographs to capture documents. The approach became commercially viable in the 1920s when New York City banker George McCarthy used specialized cameras to record paper banking records. He also developed special microfilm readers (the Checkograph) to read the records.[18]

Microfilm had many advantages over paper records. Film was more durable than paper. Film also took up a lot less space than paper and print materials. Copying microfilm was easier, cheaper, and faster than copying or reprinting a book.

When the Library of Congress sent agents to Europe and China during World War II, it sent them with microfilming cameras. Newspapers, books, pamphlets, and all manner of documents were photographed and sent back to the United States for analysis. The use of microfilm spread rapidly as a means of mass distribution of materials. Governments and newspapers alike adopted microfilm for archiving their work. By the 1950s an idea emerged: the world's knowledge could fit into a briefcase—loaded with microfilm (called microfiche when it was divided onto index card–sized sheets) and a portable film reader.

Many of the documents being shared among scholars and researchers with the growth of science and technical information discussed in chapter 6 were done on microfiche. You will still find cabinets filled with microfiche in many academic libraries.

With the advent of desktop computers, and particularly with the wide availability of CD-ROM drives, microfiche was scanned and distributed in digital format. Now researchers and journalists alike could read documents, though in the early days of CD-ROM the documents were sometimes just crude scans of black-and-white microfiche that itself was a sometimes crude photo of an original paper document. Eventually publishers and database builders like EBSCO and Elsevier skipped microfilm altogether and distributed digital documents.

However, by 2002 a vast amount of the world's books and documents were still very much physical artifacts. In 2002 a relatively new, but wildly successful, search engine company by the name of Google (founded only four years earlier in 1998) decided that this had to be changed. It launched an effort to scan the world's books.

> By 2004, Google had started scanning. In just over a decade, after making deals with Michigan, Harvard, Stanford, Oxford, the New York Public Library, and dozens of other library systems, the company, outpacing Page's prediction, had scanned about 25 million books. It cost them an estimated $400 million. It was a feat not just of technology but of logistics.[19]

I recall talking to a Harvard librarian at the time who said their number-one logistical problem was not scanning, but filling the trucks on the fabled Harvard Yard with books to be processed.

This is not to say that there weren't issues with the scanning effort. Many librarians complained about the quality of the scanning. Many more complained about putting this massive effort in the hands of a commercial company. However, the real issue that ultimately doomed the project is a lesson in yet another feature of the knowledge infrastructure: intellectual property law.

The skepticism of librarians, and even some within Google, about the company's ability to scan all the books in the world was nothing compared to the outright anger of publishers when Google tried. You will recall the discussion of the first sale doctrine in chapter 2. When you buy a physical book, you have the right to resell it, lend it out, and so on. There is some precedent for making a copy of the work as well for preservation. However, Google didn't buy these books—the libraries Google was working with did.

Google's legal justification for not seeking the permission of publishers to scan copyrighted works was the same one it had successfully made for indexing websites. Google didn't distribute the content of the websites, just links and summaries. Google didn't want to put these books online in full, but was simply scanning them to index them and make them findable through their search engines. The publishers didn't buy it. In 2011, 25 million scanned books into the project, the Authors Guild, an organization representing authors and publishers, sued Google.

The Google Books Search Amended Settlement Agreement was a 165-page document with more than a dozen appendices that took two and a half years to negotiate. The proposed settlement would have cost Google about $125 million, including a one-time $45 million payment to the copyright holders of books it had scanned along with $15.5 million in legal fees to the publishers, $30 million to the authors, and $34.5 million toward creating the registry of scanned books.[20] In the end, no settlement could solve the issues of authors, publishers, libraries, Google, and Google competitors like Microsoft and Amazon. Google did, however, win the case of *Authors Guild v. Google*, but that only allowed Google to display snippets of published books already scanned.

The entire process also raised an interesting issue directly related to all of those archives of newspaper and television news programming: It is hard to be forgotten on the internet.

FROM BURYING THE LEDE TO THE RIGHT TO BE FORGOTTEN

While authors and publishers worried about Google scanning the world's books, they were also very excited about potential new audiences for out-of-print and low-selling books. A running theme of these conversations concerned "the long tail." The idea is that only a few books sell in the millions of copies, a few more sell thousands of copies, and a lot more sell hundreds. Publishers and authors alike were keen to highlight the low-selling books in hopes of finding a new audience for their existing inventory. Search engines could change the nature of memory in the knowledge infrastructure, keeping ideas and materials alive long past the half-life of publication.

The long tail reality—making older and small-print-run materials easily findable—hit the news media in unexpected ways. Newspapers have been publishing stories online since the late 1990s, in addition to putting up archives of even older stories. As papers went online—cautiously at first, but then in search of new advertising revenue—they have made their content available to search engines and helped people find their older stories. The page views and embedded ads in the old stories are a vital part of their revenue.

Now, through search engines, anyone with an internet connection can find news articles written a decade ago without having to go to a library or document archive to scan through hundreds of yards of microfilm. For some, it has become too easy to find old stories. People accused of crimes and misconduct 20 years ago are suddenly finding these stories popping up again. With internet and social media searching becoming commonplace in job interviews, it is problematic that youthful indiscretions or painful past circumstances are as accessible as an article in today's papers. Worse still, the old articles covering bad behavior can be found with no link to later exonerations or context.

The European Union Parliament passed laws that sought to protect their citizens from the re-emergence of dated materials. The parliament inscribed into law a right to be forgotten. By law an EU citizen could request that damaging information be removed from search engines like Google. The right to be forgotten might be to leave behind long-ago criminality, or to combat "revenge porn," in which a sexual partner posts explicit content without a person's permission.

To be clear, the right to be forgotten is far from a simple matter. Courts, search engines, and news organizations have to balance this new right against the issue of freedom of speech, and a drive to document and preserve the past. In news organizations, search engines are challenging the idea that news is the first draft of history, because now the first draft, and every subsequent draft, can live side by side with equal weight.

Forgetting is a vital part of the knowledge infrastructure, because it is a vital part of knowledge. As I wrote earlier in the chapter, human memory and cultural memory are not perfect. We highlight and diminish experiences, beliefs, and facts all the time. We do this as we constantly form and shift our view of the world. Now we must face up to the fact that technology makes it harder to have control over this process. We are now facing an infrastructure that is primed to overload us with information or builds filter bubbles that shield us from potentially revelatory information. Where is the middle point?

Adding to this complexity is the fact that we are creating digital records of our everyday activities without knowing that these records are becoming available on the net. It is one thing to find an article written on your drunk driving arrest 30 years ago; it is quite another to find out that Google Maps has kept track of every business you have visited for the past five years, and you never knew. Who has access to that information? Who has the right to keep or delete those records? With so many apps keeping track of your location, your internet searches, your dating preferences, your conversations and, well, your life, how would you even know what is being remembered about you to begin with? And of course, with today's digital landscape, who owns all of that data, and who has the right to sell it?

With the right to be forgotten, the European Union looked to law to control a society's memory. The United States also looked to law to regulate how digital records were used over time, but sought to remember everything.

FROM BUNS AND COMFORTABLE SHOES TO WARRIORS FOR PRIVACY

On September 11, 2001, two planes were deliberately flown into the World Trade Center's twin towers in New York City. Terrorists crashed another commercial airliner into the Pentagon in Washington, DC, while a fourth plane crashed in the woods of Pennsylvania. In these 9/11 attacks, 19 al-Qaeda terrorists killed 2,977 victims and injured over 25,000 people, and these figures do not consider the lasting deaths and injuries experienced by first responders searching and cleaning the rubble pile in New York. The United States has been in a state of national emergency ever since.[21]

The horrors of 9/11 would drive a new wave of war and conflict abroad for Americans. It would be used as justification for the invasion of Afghanistan and Iraq. It would lead to greatly expanded military funding. 9/11 would also increase attacks on Muslim Americans and suspicion of predominantly Muslim countries.

In the weeks following the 9/11 attacks, the news media would not only report on the aftermath, but in some cases, advocate for military action. Years later, newspapers and television news operations would be accused of a sort of dereliction of duty, not critically analyzing messages and actions coming out of the White House and the US military. A wartime footing in the media promoted ideas like the War on Terror and even self-censorship about renditions, secret foreign prisons, and torture.[22] As the *Fordham Urban Law Journal* put it:

> The troubling element of the coverage was not the patriotic slant, however, but the media's decision to suppress debate over sensitive topics, like the "why's" behind the terrorist attacks, the history of U.S. policy in the Middle East, and the long-term impact of the government's new powers. Even newsman Dan Rather fell victim to the times, weeping with talk show host David Letterman a few days after the attacks and pledging: "George Bush is the President. . . . Wherever he wants me to line up, just tell me where."[23]

Another area that received little direct and critical attention was in the Patriot Act. In the days after the 9/11 attacks, staffers scrambled in Capitol Hill to hammer out new laws that would become part of the U.S. response to the attacks. One of these bills would become the USA Patriot Act.*

The Patriot Act was focused on expanding the powers of law enforcement to identify, track, and detain terrorists, domestic and foreign. It had among its principal provisions the creation of the crime of "domestic terrorism"; a diminished threshold for detention and deportation of non-citizens; increased sharing and collaboration between federal law enforcement and intelligence agencies; enhanced surveillance powers for the government in gathering foreign intelligence information, including lessening judicial oversight in wiretapping (and expanding the allowable wiretapping to include roving wiretaps not tied to a single telephone number); new scrutiny of banking; as well as provisions to expand border security.

---

* The USA Patriot Act, or simply the Patriot Act, is technically an acronym of Uniting and Strengthening America by Providing Appropriate Tools Required to Intercept and Obstruct Terrorism Act of 2001.

In the days following the 9/11 attacks there was one group that joined these discussions in spite of the fear of another attack: the American Library Association. The reason? Because the Patriot Act was not written from scratch, but rather was seen as a way of pushing through a number of domestic surveillance measures that had been discussed for years. The librarians of the ALA felt that these measures went too far in terms of violating citizen privacy and civil liberties.

When the Patriot Act was signed into law 45 days after the attacks by a vote of 357–66 in the House and only one opposing vote in the Senate, it mobilized librarians once again. However, this would be a very different type of campaign compared to that of World War I and World War II. Rather than simply supporting the troops through reading, or engaging in intelligence operations, American librarians would team with other organizations such as the American Civil Liberties Union (ACLU) to fight the Patriot Act. The primary provision the librarians set their eyes on was Section 215.

Section 215 required institutions, public and private, to turn over records concerning an individual to law enforcement. Government law enforcement and investigators have long been able to get access to these kinds of records. However, the Patriot Act made two major changes in how they did it. The first was that the FBI or another government agency didn't need a judicial warrant. Rather than going before a judge, the agency could provide a National Security Letter that had the power of a subpoena. The second was that all such record requests came with a gag order—legally, the contacted institution couldn't tell anyone that it had been contacted or what records had been shared. Therefore, the person under surveillance or the institution receiving the request could not go to court to challenge the request.

Section 215 became known as the "library provision" because librarians were vocal in their opposition to it. The argument from librarians was that such secret surveillance threatened the privacy of their patrons. If no one knew whether the FBI was looking at their history of books borrowed, or web sites accessed on a public computer, then folks would self-censor.

Here, once again, librarians were being recruited into a conflict—the War on Terror—but in a radically different way than President Roosevelt encouraged librarians to the government side in World War II when he declared:

> Libraries are directly and immediately involved in the conflict which divides our world, and for two reasons: first, because they are essential to the function-

> ing of a democratic society; second, because the contemporary conflict touches the integrity of scholarship, the freedom of the mind, and even the survival of culture, and libraries are the great tools of scholarship, the great repository of culture, and the great symbols of the freedom of the mind.[24]

Section 215, the argument goes, is trampling on the freedom of the mind.

As in World War II, librarians took action. Librarian Jessamyn West encouraged all libraries to hang a sign stating "The FBI has not been here. (Look very closely for the removal of this sign,)"[25] The ALA put out policy statements, and in 2005, four Connecticut librarians sued the federal government over a National Security Letter they had received. A lawsuit may not seem like an act of bravery, but losing the lawsuit meant almost certain imprisonment for the librarians. They were directly violating the gag order and breaking national security law.

> In 2005 four librarians from Connecticut went to court to challenge a National Security Letter demanding patron data. National Security Letters, whose use was expanded in the Patriot Act, allow the FBI to demand massive amounts of user data from Internet providers without a warrant or judicial review, and they're always accompanied with a strict gag order. Both the court case and the National Security Letter were eventually dropped, and the Connecticut librarians are among the only recipients of a surveillance gag order who can speak openly about their experiences. That same year, the American Library Association—which has listed intellectual freedom and privacy as core values since 1939—filed a brief in the Supreme Court to challenge the Patriot Act.[26]

I would just like to take a moment to point out that the stereotype of the quiet and retiring librarian, a creation of mass media, could not be further from the truth. Librarians have always been active agents of change in the knowledge infrastructure. Social action on the Patriot Act and the Foreign Service of the Library of Congress described in this chapter are only two examples.

Librarians were active agents of change and freedom of information in other conflict zones as well. Abdel Kader Haidara is a librarian from Timbuktu who helped safeguard the printed treasures of his city against nearly 1,000 Islamist fighters from one of al Qaeda's African affiliates who occupied his city, al Qaeda of the Islamic Maghreb.[27] Librarians of Alexandria supported the protestors during the Arab Spring of 2011,[28] Alia Muhammad Baker, the chief librarian in the Al Basrah Central Library, heroically saved

30,000 books from destruction during the Iraq War.[29] And Yvonne Cech, a librarian at Sandy Hook Elementary School, protected 18 fourth-grade children and three staff from a school shooter by locking them in a storage closet and barricading the door.[30]

Throughout this text I keep emphasizing that it is people and their passions that matter, and shape the knowledge infrastructure with their actions and choices. No amount of data can convince a person to risk death to preserve their culture, their gift to the next generation. To the stories of Haidara, Baker, and the librarians of Connecticut I could add countless reporters who enter war zones to seek the truth, or French and Dutch citizens who risked their lives to provide intelligence on the Nazis to the allies. We must remember that war has crafted the way we see and learn about the world in positive as well as negative ways.

I would also like to point out that the actions of these individuals often have ripple effects they are unaware of. While Section 215 of the Patriot Act *was* known as the library provision, you are likely much more familiar with the consequences of the act from another name: Edward Snowden. The very same provision that allowed the government to collect library records was used to justify the mass collection of data from phone companies and internet service providers.

Up to this point I have been talking about how commercial interests like advertisers, retailers, and tech companies collect and use your digital trail on the internet. With the Patriot Act, the government entered the game in a very big way. Arguing that it only collected metadata about phone calls, the government installed systems at major phone carriers to scoop up information on all the calls coming into, within, and going out of the country. While doing so, the government installed systems to inspect internet packets, and where they were going as well.

Don't let the term *metadata* fool you; it is meant to sound both obscure and unimportant. The metadata of a call is not the content of the call—a recording of what is being said—rather, it is the phone numbers of all the parties in the call. And it's the time the call lasted and the location of the callers. If that sounds unimportant, let me give you a scenario. You are having an affair.* Do I need to know what was said in order to infer who is part of

* Shame on you, but it's not my place to judge.

the affair by the phone numbers you call and receive (and when and where)? Also, if any of the calls came from or to a foreign number, then the government can record the content of the call. Recall that this gathering of records also includes your web traffic. Are you ready to publish your search history (in open and private mode) to the world? I hope so, because that is the activity Edward Snowden revealed was happening within the National Security Administration and beyond.

FROM DEPOSITORIES TO DIGITAL DARK AGES

This trip through history, focusing on memory organizations and the repositories of media, have highlighted a central problem in shaping the knowledge infrastructure. There are two competing, and seemingly opposite, requirements: to remember (the preservation of documents in World War II, the importance of cultural heritage, microfilming, and mass digitization), and to forget (privacy and Section 215 of the Patriot Act, and the EU's right to be forgotten). This seeming battle is more relevant today than ever before. Just take President Donald Trump's tweeting.

In 2017 the Knight First Amendment Institute sued the President Trump for how he used Twitter.[31] There were two key issues: the first was whether Trump tweeting as @realDonaldTrump constituted an official presidential account and, as such, could block other Twitter users. The courts said the account could not. The second question (as of this writing undecided) was whether Trump could delete previous tweets. You see, there are laws in place that control the life of a presidential document. These documents are not supposed to be destroyed, but rather turned over to the National Archives for historical, and in more modern times, for legal review. If tweets are presidential records, deleting them could be against the law.[32]

It is not just the office of the president. Most federal agencies are required to keep ongoing records of their publications and disseminations. The Government Printing Office was formed in 1860 and renamed the Government Publishing Office in 2014 as a means to capture the output of government for review. The GPO was an attempt to save costs and provided printing services for all federal agencies. In 1895 the law authorizing the GPO was changed to require agencies to not only use the GPO but make the printed documents available through the newly formed Federal Depository Library Program. The depository program sent copies of important government publications to libraries throughout the country so that citizens could keep an eye on what the government was doing.

Today there are over 1,100 depository libraries throughout the United States, most at academic and public libraries. These documents have been primarily stored as microfiche since the 1940s. However, as the internet became more widely available, many agencies started distributing information, publications, and other materials exclusively online. The reasoning was twofold. For the optimists among you, it was to allow for cheaper distribution on computers, then laptops, then phones of citizens everywhere. For the cynic, it was a way to bypass the depository program, and thus the archiving requirements (for a time at least) for government documents. Instead of documents being fixed in plastic film, the website could shift and change as the political winds did. To be fair, the Federal Records Act and the E-Government Act of 2002 place some obligations on the executive branch to archive certain web-only materials.[33] Let's just say that the interpretation of required, recommended, and best practices in this area is fluid.

The federal depository program is not unique to the United States. Most nations have these kinds of obligations. The Nazis did in World War II . . . forming the documentary evidence that both convicted German officers at the Nuremberg trials for war crimes and fed the Marshall Plan in building democratic governments in Europe after the end of hostilities. The ability to capture immense volumes of data on citizens and customers allows for great advantages in pursuing terrorists and in surveilling and detaining citizens. As our tools to collect and analyze this data increase in sophistication while simultaneously decreasing our ability to understand their internal operations (see the discussion of deep learning in chapter 5), are we comfortable that, as individuals and as a society, we have found the right mix of being remembered, and being forgotten?

NOTES

1. John Keegan, *The First World War* (New York: Alfred A. Knopf, 1999).

2. Wayne A. Wiegand, "The Library War Service: ALA's Book Campaigns in World War I," *American Libraries*, February 18, 2016, https://americanlibrariesmagazine.org/2016/02/18/ala-history-library-war-service.

3. Richard Allen Greene, "King Leopold II Statues Are Being Removed in Belgium. Who Was He?" CNN, last modified June 11, 2020, https://www.cnn.com/2020/06/10/europe/belgium-king-leopold-ii-statue-intl/index.html.

4. "War Service Library Book," American Library Association, April 19, 2010, http://www.ala.org/tools/war-service-library-book.

5. "History," American Library in Paris, accessed July 22, 2020, https://americanlibraryinparis.org/history.

6. Frederick J. Stielow, "Librarian Warriors and Rapprochement: Carl Milam, Archibald MacLeish, and World War II," *Libraries & Culture* 25, no. 4 (1990): 513–33.

7. Ibid.

8. Not all librarians were in support of the war effort; the Progressive Library Council, started in 1939, actively opposed involvement in the war. See Rosalee McReynolds, "The Progressive Librarians Council and its Founders," *Progressive Librarian* 2, (1990/1991): 23–29, http://www.progressivelibrariansguild.org/pdf/mcreynolds.pdf.

9. Davide Rampello, *Human Capital: A Journey Through the Italy of the G7* (Milan: Skira Editore, 2017), n.p.

10. "World Heritage List," UNESCO (United Nations Educational, Scientific, and Cultural Organization), accessed July 22, 2020, https://whc.unesco.org/en/list.

11. "The Organization's History," UNESCO, accessed July 22, 2020. http://www.unesco.org/new/en/%20unesco/about-us/who-we-are/history.

12. "Convention Concerning the Protection of the World Cultural and Natural Heritage," UNESCO, accessed July 22, 2020, https://whc.unesco.org/en/conventiontext.

13. Ibid.

14. Ibid.

15. Stephennie Mulder, "Trump's Threats Against Iranian Cultural Sites May Unite Iran More Than Soleimani's Death," *USA Today,* January 8, 2020, https://www.usatoday.com/story/opinion/2020/01/08/iran-cultural-sites-trump-qassim-suleimani-column/2832522001.

16. Courtney Subramanian, "Donald Trump Again Threatens to Target Iranian Cultural Sites Amid Mounting Tensions Over Qasem Soleimani Killing," *USA Today,* last modified January 7, 2020, https://www.usatoday.com/story/news/politics/2020/01/05/iran-trump-mocks-congress-war-powers-act-notification-over-action/2818355001.

17. "Global Strategy," UNESCO, accessed July 22, 2020, https://whc.unesco.org/en/globalstrategy.

18. "Brief History of Microfilm," Microfilm World, accessed July 23, 2020, https://www.microfilmworld.com/briefhistoryofmicrofilm.aspx.

19. James Somers, "Torching the Modern-Day Library of Alexandria," *Atlantic,* April 20, 2017, https://www.theatlantic.com/technology/archive/2017/04/the-tragedy-of-google-books/523320.

20. Ibid.

21. Donald J. Trump, "Text of a Notice on the Continuation of the National Emergency with Respect to Certain Terrorist Attacks," the White House, September 12, 2019, https://www.whitehouse.gov/briefings-statements/text-notice-continuation-national-emergency-respect-certain-terrorist-attacks.

22. Raymond Bonner, "The Media and 9/11: How We Did," *Atlantic,* September 9, 2011, https://www.theatlantic.com/national/archive/2011/09/the-media-and-9-11-how-we-did/244818.

23. Lisa Finnegan Abdolian and Harold Takooshian, "The USA Patriot Act: Civil Liberties, the Media, and Public Opinion," *Fordham Urban Law Journal* 30, no. 4 (2003): 1429–53, https://ir.lawnet.fordham.edu/cgi/viewcontent.cgi?article=2097&context=ulj.

24. Cited in Frederick J. Stielow, "Librarian Warriors and Rapprochement: Carl Milam, Archibald MacLeish, and World War II," *Libraries & Culture* 25, no. 4 (1990): 513–33.

25. April Glaser, "Long Before Snowden, Librarians Were Anti-Surveillance Heroes," *Slate,* June 3, 2015, https://slate.com/technology/2015/06/usa-freedom-act-before-snowden-librarians-were-the-anti-surveillance-heroes.html.

26. Ibid.

27. Joshua Hammer, "The Librarian Who Saved Timbuktu's Cultural Treasures From al Qaeda," *Wall Street Journal*, April 15, 2016, https://www.wsj.com/articles/the-librarian-who-saved-timbuktus-cultural-treasures-from-al-qaeda-1460729998.

28. R. David Lankes, *Expect More: Demanding Better Libraries for Today's Complex World* (Jamesville, NY: Riland Publishing, 2016).

29. Shaila K. Dewan, "After the War: The Librarian, Books Spirited to Safety Before Iraq Library Fire," *New York Times*, July 27, 2003, https://www.nytimes.com/2003/07/27/world/after-the-war-the-librarian-books-spirited-to-safety-before-iraq-library-fire.html.

30. Nikki DeMarco, "Heroic Librarians: Unexpected Roles and Amazing Feats of Librarianship," Book Riot, February 12, 2020, https://bookriot.com/heroic-librarians.

31. "*Knight First Amendment Institute v. Trump*," *Wikipedia*, last modified July 23, 2020, https://en.wikipedia.org/w/index.php?title=Knight_First_Amendment_Institute_v._Trump&oldid=918930368.

32. Though that is far from a settled question. as seen in John Kruzel, "What Does the Law Say About Donald Trump's Deleted Tweets?" PolitiFact, September 27, 2017, https://www.politifact.com/article/2017/sep/27/what-does-law-say-about-donald-trumps-deleted-twee.

33. "NARA Guidance on Managing Web Records Background," National Archives, last modified August 15, 2016, https://www.archives.gov/records-mgmt/policy/managing-web-records-background.html.

9

Media Consolidation

From Steamboat Willie to Disney+

The cutting of Germany's submarine telegraph cables in August 1914 was an act of power. It was a show of force from a military standpoint, a demonstration of British resolve and strategy. It was also an act of physical power. If you go back and re-read the first three paragraphs of this book, you will see a brief mention that the *Alert* "struggled" to cut the fifth and last telegraph cable connecting Germany to the rest of the world. I'd like to take a moment to give a little more depth—excuse the pun—to what "struggled" meant.

The laying of submarine telegraph cables between Great Britain and Canada was first attempted in 1857—it failed. It was attempted again in 1858—it failed. It wasn't until the third attempt, with two ships starting together in the mid-Atlantic and then heading away from each other to the east and the west, that the link was made. However, after parades and rejoicing on both sides of the Atlantic, the cable could only send a single character every 3 to 8 minutes or so, and within two months the insulation shielding the cable disintegrated under high voltage, and the cable failed.

In your mind you may have an image of a cable that is somewhat thin and flexible, like the telephone lines strung up next to highways. That was indeed the first set of attempts—thin and flexible. Those terms are relative, however. The first cables consisted of long lines of copper insulated with a natural occurring latex from India and covered in tarred hemp, then sheathed in iron wire. These cables were "thin," but still weighed over a ton per mile. The cables that eventually would successfully make the span, leading to the mission of the *Alert*? They weighed twice as much. The *Alert* was struggling to pull literally tons of cables from the sea floor under the prospect of imminent attack from an approaching destroyer group.

The *Alert* itself represented a third, very different kind of power: the power in controlling ideas. The *Alert* was a cable ship built just for the purpose of laying down submarine cables. The British government, major shippers, and budding telegraph companies invested, in today's terms, millions of dollars (well, pounds) in the effort to expand the global telegraphy network. It was important to all involved that the effort be protected,[1] and that the power of telegraphy, as much as possible, stay in the hands of the British—as much a concern for national security as industrial success.

In the United Kingdom, this was done through patents. Patents represented government protection of inventions (machines and processes). When I say that the British controlled the telegraphy network during World War I, much of this control came from patents that prevented other nations and companies from competing. The idea behind a patent, then and now, is twofold: give an inventor a limited monopoly over the device or processes, to either make money or control the use of the invention; and make the invention widely available to the public after the limited monopoly. In essence, reward smart people in the short term, but ultimately benefit everyone with their cleverness.

A current example is the development and sale of new medications. Pfizer spent a great deal of money developing the drug sold under the brand name Lipitor, to control cholesterol. This included the basic chemical science and a series of animal and human clinical trials to ensure the efficacy and safety of the drug. In return for this investment, the U.S. Patent Office extended to Pfizer the exclusive right to sell the drug at whatever price they wish for 20 years.

Now, that is 20 years from the point of patent, normally before the clinical trials, so in effect Pfizer may have had 10 or fewer years of exclusive access to the market,[2] After the patent period expires, any company could make and sell the drug. And in fact, the patent itself acted as a blueprint on how to manufacture it, meaning other companies quickly created generic versions, and competition drove down the price. Pfizer got the initial reward, but everyone with high cholesterol benefits in the long term because the knowledge has been shared.

Patents are only one form of protection given to ideas, inventions, expressions, and other forms of intellectual property in the United Kingdom, the United States, and globally. Where patents protect inventions, copyright protects *expressions*; that is, the tangible form of creativity (a book, a poem, a song, or even the choreography for a dance). Folks often think that copyright protects ideas, but it doesn't—just the words, images, or recordings of the ideas. So, this book has a lot of ideas about "knowledge infrastructure," but only the arrangements of the words I use to express that idea are protected. You can't copy the book, but you can (and should) use the ideas. Write your own book or article, or make a music video about the concept of the knowledge infrastructure.

Once again, as in patents, the goal with copyright is twofold: benefit the creator, but ultimately add to the public good. And, as we'll see in this chapter, it is frankly a better idea than the reality has become. For example, when Walt Disney first copyrighted Mickey Mouse in 1928 as part of the film *Steamboat Willie*, Disney could expect his exclusive rights to Mickey to last for 56 years, after which time the movie and the characters would enter the public domain where anyone could use them, remix them, copy them, and more. Through legislative extensions we'll discuss later, this term is now up to 120 years.

Trademarks are often also included as intellectual property protection, but these are slightly different. A trademark is an instrument of doing business (though not necessarily for profit) that is meant to prevent marketplace confusion. Trademarks prevent five burger joints from all claiming the name McDonalds and then providing very different quality or experiences. So a trademark is intellectual property protection, but with a different goal.

As I continue the trips through time, the power of owning media will become increasingly important. If the knowledge infrastructure is about making people smarter, the policies controlling what ideas and expressions we have access to and can benefit from are crucial. In this chapter I am going to explore the concept of power through control of the media. And I am going to start by looking at another channel of mass media—television—and how it was initially shaped by the use of patents; used the power of programming market share to both stop a war and forever change how war was seen through the media; and finally, how the consolidation of intellectual property affects our view of the world. To do all of that, I have to start with a spinning disk and a farmer from Utah.

FROM RADIO STAR TO VIDEO KILLING THE RADIO STAR

Television was around during World War I, but as a series of experiments to further the capabilities of radio. At about the same time that Marconi was trying to capitalize on Alexander Graham Bell's development of a working microphone to encode sound into electrical signals for his radio service, others were trying to encode images.

Simply sending images via radio waves was accomplished in 1843 by Scotsman Alexander Brin. By 1856, Giovanni Caselli had a working facsimile service in Italy. In 1884, Paul Gottlieb Nipkow patented the "Nipkow disk," which used a spinning disk with a series of spiral holes to scan images into electrical signals. In 1907 the refined development of vacuum tubes made the disk a practical system for capturing and transmitting images. This mechanical drum system would be used in the first television cameras through World War I. By 1911, the development of the cathode ray tube (CRT) filled in the receiving end of the television equation.

By 1925 crude moving images of faces could be sent over radio waves, and 48-line images were being sent over a span of miles. A patent was awarded to Charles Jenkins for the technology. In Japan in1927, Kenjiro Takayanagi had managed to transmit 100 lines from a mechanical disk camera to a CRT. Also in 1927, American farmer and inventor Philo Farnsworth received a patent for his "image dissector" that acted as an all electronics–based practical television camera. In 1935, Vladimir Zworykin was also awarded a patent for an all-electronic system that replaced the mechanical spinning disk in capturing images with an "iconoscope." These two patents—one by Zworykin and one by Farnsworth—would lead to one of the early patent wars.

Zworykin worked for RCA, the dominant radio company founded in part by Marconi. They sought to invalidate Farnsworth's earlier patent, claiming it was too broad. The Patent Office sided with Farnsworth, and in 1939 RCA licensed Farnsworth's invention for $1 million to be paid out over 10 years.

Farnsworth was prepared to be a very rich man, believing, correctly, that commercial television would take off. The problem for him was that on September 1, 1939, the Nazis invaded Poland. World War II would freeze development of television as a technology and a business. By the time the war ended, Farnsworth's patent was set to expire and there simply wasn't enough time left to reap the benefits of the television revolution to come. The delay caused by the fighting patents and the war had killed his dreams.

By 1945 television was a mixed bag of services and technologies. Europe, for example, had adopted a 625-line standard (the resolution of the picture being defined by how many lines could be projected vertically) developed in Russia, while the United States had settled on 525 lines. Even then, television being a radio-broadcast system, different areas within the United States might still have different technologies for transmitting and receiving broadcasts. However, standards were soon put in place, and radio companies used their existing broadcast infrastructure and tidy profits to venture into the world of television.

In 1946 there were approximately 6,000 TV sets in the United States. By 1951, that number had grown to more than 12 million.[3] By 2011, 96.7 percent of American households owned at least one TV.[4]

FROM A PATENT WAR TO A RATINGS WAR

The television industry grew around the advertising business model that began with the newspaper industry in 1830s with the so-called "penny press." Before 1830, newspaper circulation was limited to subscribers of means and focused almost exclusively on issues of business, policy, and foreign affairs. Only the well-to-do could afford the cost of a paper's production. In 1833 Benjamin Day, seeking to save his failing publishing business, launched a new daily paper called the *Sun* in New York City. Unlike other papers, Day ran articles on the interests (often sensationalized) of the common worker. He also slashed the prices of the paper using new technology, but also by adopting an advertising-driven business model. Day's paper could be cheap because advertisers were paying most of the costs in order to bring more attention to their wares. This ad-driven model drove more and more papers to seek out more and more base and brash headlines and stories to attract more readers to thereby attract more advertisers. The tabloids and clickbait of today have a direct lineage to the penny press of the 1800s.

However, unlike the newspaper industry, television adopted the network system pioneered in radio. From the end of World War II until the cable industry gained prominence in the 1980s, three television networks dominated programing and broadcast news: the American Broadcasting Company (ABC), the National Broadcasting Company (NBC), and the Columbia Broadcasting System (CBS). It is here that we see the power of access to ideas.

The 25-year period from the 1950s to the 1980s saw a great deal of power and influence invested in these three companies. This is no more evident than in the rise of the nightly news. The first national news program in the United States was the Camel News Caravan in 1949, anchored by John Cameron Swayze. It was sponsored by Camel cigarettes, which required a lit cigarette to be visible during the entire broadcast, an early demonstration of the ongoing relationship between television, news, and advertisers.

The power of advertisers in television news could be seen in Swayze's nightly sign-off: "This is John Cameron Swayze saying good night for Camel, the cigarette that gives you more pure pleasure because no other cigarette is so rich tasting yet so mild as Camel."[5] This power also directly affected the content of that news. When NBC wanted to run special reports on the Korean War in 1950, it was the reluctance of advertisers that killed the idea.

Smoking reference aside, the nightly news and the new concept of the anchor man had a strong influence on the nation. Mike Conway, author of *The Origins of Television News in America*, points out that "the networks also started to realize how viewers at home personally connected to those like Swayze who were coming into their living rooms every evening, and that this sort of familiarity and intimacy not only bred trust but also an incredibly loyal viewership."[6]

Swayze was followed in the anchor role by David Brinkley and Chet Huntley and, at CBS, by the legendary Walter Cronkite. It is worth noting in our era of suspicion of the news media that:

> When legendary anchorman Walter Cronkite retired from CBS News in 1981, a whopping 86.9 percent of Americans, according to a Harris poll, thought he could be trusted to present a "balanced treatment of the news." How things have changed. According to a recent Hollywood Reporter/Morning Consult survey, only 31 percent of Americans trust the major television news networks "a lot," with NBC's Lester Holt registering as the most-trusted TV news personality, being trusted "a lot" by just 32 percent of respondents.[7]

This article was published in February 2019.

The trust in Cronkite by the American people, a matter we'll revisit in the next chapter, demonstrates the power of the news media. It also illustrates the impact the knowledge infrastructure can have on war, and not just the other way around. In February 1968 Cronkite did a series of reports from the field in Vietnam following the Tet Offensive. Upon his return he concluded:

> To say that we are closer to victory today is to believe, in the face of the evidence, the optimists who have been wrong in the past. To suggest we are on the edge of defeat is to yield to unreasonable pessimism. To say that we are mired in stalemate seems the only realistic, yet unsatisfactory, conclusion. On the off chance that military and political analysts are right, in the next few months we must test the enemy's intentions, in case this is indeed his last big gasp before negotiations. But it is increasingly clear to this reporter that the only rational way out then will be to negotiate, not as victors, but as an honorable people who lived up to their pledge to defend democracy, and did the best they could.[8]

Many have credited Cronkite's reporting with President Lyndon Johnson's decision not to seek re-election, and the beginning of withdrawing U.S. forces in Vietnam. It is an example of how the relationship between data, media, and warfare is not one way—media can shape war. We see this, for example, in the Negro press.

Written by and for the Black community starting as early as 1827 with the paper *Freedom's Journal*,[9] the Negro press, as it is known, served a vital function in connecting and ultimately mobilizing the Black community. The articles and voice of these papers were neither neutral nor necessarily sensationalized just to attract readers. The Negro press consisted of thousands of papers serving millions of African Americans from enslavement until the end of the twentieth century. These papers served as platforms for influential Black thinkers like Booker T. Washington, W. E. B. Du Bois, A. Philip Randolph, and Marcus Garvey.[10]

The Black press was instrumental in gaining the participation of African Americans in World War I. Operated in near-isolation from the mainstream white newspapers, Black papers would be called upon to again ensure African American support of World War II when the Black community was torn between the opportunities of the war (economic and patriotic) and supporting a clearly racist society. This led to the Double V campaign, whereby strong Black editors and reporters called for victory against fascism abroad, and victory against racial inequity at home.

The thesis of this book is that the knowledge infrastructure has been shaped by conflict over the past century. To this point, those conflicts have been military. It is in the pages of the Black press that civil conflict clearly shows its imprint on today. The civil rights movement of the 1950s and 1960s seen from the white press like the *New York Times* may seem to have been a sudden eruption of racial conflict and advocacy, but the Black press shows it to be a continuation of a Black liberation movement begun in the days of enslavement and continuing to the Black Lives Matter movement of today.

Cronkite's reporting, the role of the Negro press in both World Wars, and the civil rights movement had a massive impact on the relationship between modern warfare and the media. In 1991 when the United States led a coalition of 34 countries to liberate Kuwait from Iraq in Operation Desert Storm, the ghost of Walter Cronkite was front and center (not literally, mind you; Walter Cronkite didn't die until 2009). Whereas some have called Vietnam the first televised war, Desert Storm might be described as the first war staged for television. Real-time access to satellite images and night-vision cameras mounted to the nose of missiles to transmit images reminiscent of video games fed a newly birthed 24-hour news cycle in CNN.

Unlike Cronkite's expansive and self-directed reporting in Vietnam, reporters were pooled, and the military fed the pool with the images and stories they chose. Much like the propaganda campaigns of the British in World War I, the United States and allied militaries sought to limit the information available to reporters and, thus, control the narrative. To be clear, there were exceptions to the limited coverage. CNN reporters risked their lives staying in Baghdad during the opening salvos of the offensive, and reported live and unedited.

However, just as with 9/11, many have criticized the media in Desert Storm for promoting narratives based more on patriotism than critical analysis. Of course, the entire offensive took just less than two months, leaving little time for detailed reflection, and CNN's Peter Arnett in particular broke stories of prisoners of war and civilian casualties (leading some to brand Arnett as unpatriotic).

Once again, Desert Storm showed the power of controlling access to images and narratives. It would be a blueprint attempted post-9/11 with the invasions of Iraq and Afghanistan. In both of these cases, however, time provided to be the Achilles heel of propaganda: time for reflection, to establish information channels that bypassed traditional gatekeepers, and time for public sentiment to shift.

FROM HAWKEYE PIERCE TO HAWKEYE THE AVENGER

Shaping public sentiment was not restricted to the news, however. One of the biggest events in television history was the final episode of *M*A*S*H* in 1983. The television comedy aired on CBS from 1972 to 1983. The series itself was a television spinoff from a 1970 movie of the same name, which itself was an adaptation of the 1968 Richard Hooker novel, *MASH: A Novel About Three Army Doctors*. The show was about a mobile medical hospital unit during the Korean War . . . ironic, because the series actually lasted eight years longer than the war it dramatized.

What is remarkable about the series at the time was that the book and movie were clear anti-war statements, using the Korean war as a stand-in for dark commentary against the ongoing Vietnam War. While the television adaptation was clearly gentler in its treatment, the anti-war message was a constant—scenes of trauma surgery, early treatment of post-traumatic stress disorder, sympathetic views of native "Koreans,"* all interspersed with humor and character development. Even the show's theme song demonstrates the compromises for a general audience mixed with messages of protest. Most Americans can still hum it, but few know it is the instrumental version of "Suicide Is Painless."

In a way, the final episode of *M*A*S*H* was the end of an era. With the advent of cable networks and the fracturing of the media landscape, it would become harder and harder to draw a single large live-event audience. The only larger audiences for television events in the United States are the Super Bowls from 2010 through 2017. Other network series would draw high ratings, such as *Cheers* and *Seinfeld*, and with the introduction of syndication, older series like *I Love Lucy* and *Star Trek* would draw massive viewership. But it just wouldn't happen live and in real time. This lack of a real-time mass audience has an impact on the idea of a national narrative. Narratives, their complexity, and the access people have to them matter.

In her TED talk,[11] writer Chimamanda Adichie stressed the importance of breaking down stereotypes and prejudices of foreign cultures through literature. She contends that for many in western cultures, places like Africa are seen in shallow two-dimensional terms because we have so few narratives covered by our media. She talks about how many in the west see

* At the time, most Koreans were portrayed by Japanese American actors.

African countries like Nigeria as places of poverty, or sickness, or civil strife because those are the only stories we encounter. In Nigeria, as in all cultures and countries, there are stories about growing up, about crime, about love, and about ambition. Only by opening ourselves up to conversations beyond simple event-driven news can we truly come to empathize with a people. The power of narratives is a recurrent theme in the international women's movement. Keeping girls from attending school and restricting access to literature and information about cultures where women have power are seen as means of oppression.

Much of this oppression comes in the form of direct censorship. China, for example, has built an extensive and sophisticated technical system for censoring ideas and dissent. Known as the Great Firewall of China (or GFW), the system of laws, technology, and censors arrayed to control Chinese citizens' access to information stands as the starkest challenge to the idea that the internet and media are liberating forces.

The GFW doesn't simply block open internet sites like Facebook and Google. It actually searches out the information available to the Chinese people and can identify potential threats to government ideology at the word level. In China there are phrases and terms that simply are not permitted on the internet. News stories from the outbreak of the novel coronavirus to government corruption cases are wiped from the internet, or replaced with tightly controlled official information. Where many see the internet as a liberating force, China has proven it can be an extremely effective means of narrative control.

In China, technology and media are power, and retaining centralized control of it has allowed the country to distribute control of the economy while retaining centralized political power. China has catapulted to the second-largest economy in the world by shifting from the old communist centralized market system to a capitalist economy of private production (highly regulated and subsidized by the government). When this transition occurred in most former Soviet countries, political and communications power also shifted to democratic participation—a trend currently in great peril with the rise of far-right and nationalist movements in Europe and the former Soviet republics.

Some would argue that this trend away from democracy and toward a more controlled media is happening in the United States as well. Clearly a shift in narrative power is occurring. However, this shift neither started with the election of Donald Trump, nor is it limited to nationalism, far-right ideologies, or politics. Over the past three decades, even before the rise of the internet in everyday life, the United States began shifting from a distributed, though somewhat chaotic, knowledge infrastructure to one increasingly dominated by a few powerful players and walled off behind paywalls.

Part of the thinking behind the creation of the Federal Communications Commission in 1934 was to regulate the ownership of television and radio stations to ensure local voices on the airwaves. In 1934 there were 583 radio stations in the country[12] and, for all intents and purposes, no television stations. By 1983 there at least 6,300 AM and FM stations[13] and over 1,000 broadcast television stations.[14] By 2012, there were 11,336 radio and 1,781 TV stations.

While FCC regulation has varied in both its definition of "promot[ing] localism and competition" and its enforcement, station ownership consolidation is increasing:

> In 2004, the five largest companies in local TV—Sinclair, Nexstar, Gray, Tegna and Tribune—owned, operated or serviced 179 full-power stations, according to a Pew Research Center analysis of Securities and Exchange Commission filings data. That number grew to 378 in 2014 and to 443 in 2016.[15]

What is driving this consolidation? In part, companies are seeking greater ad revenue including from a near tidal wave of political ad buying unleashed by the 2010 U.S. Supreme Court *Citizens United* decision. That case eliminated a whole host of restrictions on political spending by corporations and the wealthy: "In 2016, the five biggest local TV companies' total broadcasting revenue was $8.3 billion, or 30% higher than in 2014, the previous major election year. And political advertising revenue grew by 31% during this two-year period."[16]

This consolidation of broadcast outlets has also supported corporate political influence. The Sinclair Broadcast Group, now the largest owner of local television stations, is a clear and declared supporter of conservative issues.[17] While Sinclair Broadcast Group stations are affiliated with NBC, CBS, ABC, and FOX, the news on all of their stations reflect a conservative political bias—even though this tends to lead to a smaller audience for the stations' newscasts.[18] At these stations local news stories tend to be displaced by national ones, and nationally produced commentaries are distributed and presented as part of local newscasts.

Behind the consolidation of station ownership lies an even more striking set of statistics. In 1983, 90 percent of U.S. media was controlled by 50 companies; in 2011, 90 percent was controlled by just 6 companies: General Electric, News Corp, Disney, Viacom, Time Warner, and CBS.[19] These companies owned 70 percent of the cable content compared to the 3,762 businesses that control the other 30 percent. News Corp, aside from owning TV stations, also owned the top newspapers on three continents (North America, Europe, and Australia).

Since 2012, when *Business Insider* provided those statistics, even more consolidation has occurred. The Disney Company, for example, has since purchased the entertainment arm of News Corp including the 21st Century Fox movie studio and catalog of films. CBS bought Viacom. In 2011, 90 percent of media was controlled by six companies—in 2019, it is down to four: Comcast, Disney, ViacomCBS, and AT&T (which was Time Warner).

These corporations represent unprecedented power in today's media landscape—power through their reach with newspapers, television stations, film studios, and radio stations. They also represent a different kind of power through their control of intellectual property. These agencies own the copyright to an enormous portion of the media that Americans consume. Disney alone owns the rights to the top-grossing film franchises of Star Wars, Pixar, Disney's own films, and Marvel. The company also owns major television and cable properties with ABC, ESPN, and FX.

These are valuable intellectual properties to Disney, and as British telegraphy companies staunchly defended their intellectual property through patents, today's media conglomerates defend theirs through copyright. Machine learning software scans sites like YouTube and Facebook looking for unauthorized use of their brands and characters. The software, upon finding such use, issues a takedown notice asking sites to remove the material, even if it is clearly a use permitted by law. To understand exactly what that means, first I need to explain the number 512.

FROM SAFE HARBORS TO AUTOMATED TAKEDOWNS

The last time the copyright laws in the United States were updated was 1998. The bill was named the Digital Millennium Copyright Act (DCMA) and it had several purposes and several intended (and many more unintended) effects. The first thing it did was bring U.S. copyright law into alignment with international treaties. The DCMA also barred any attempts or tools used to circumvent copyright. This is the exact legislation that allows for digital rights management discussed in chapter 2 for tractors and Teslas, and outlaws attempts to bypass copyright with software like Napster.

The other thing the law sought to do was be proactive in terms of the growing internet reality. One common misconception about copyright is that in order to protect something (a song, a book, a film), you had to register the piece with the Copyright Office at the Library of Congress. However, ever since 1989, copyright is automatically granted at the point of creation. That is, if you draw an amazing picture, it is protected under copyright. No need to register it, or even scribble the famous © on it. The only reason to register a copyright is as a sort of proof and protection if someone claims you copied their work.

This means that now, as it did in 1998, a lot more things that were being shared online were actually protected by copyright. Websites that allowed people to post materials, from photos on Flickr to lesson plans at the Department of Education to videos on YouTube, were feeling uncomfortable. They were afraid that one of their users would post a copyrighted work and the website would be sued.

In order to avoid this, the DMCA has Section 512, better known as the Safe Harbor provision. This provision more or less states that if someone posts an infringing piece of content, the web host can't be sued for copyright infringement so long as the host has a procedure for responding to and acting on complaints by copyright holders. So, if I post the whole *Toy Story* movie to YouTube, Disney can't sue YouTube. That is, if YouTube has a procedure in place for Disney (or anyone who feels their copyright has been violated) to claim infringement and be guaranteed a response, such as taking down the offending video and/or deleting the account of repeat offenders.

This safe harbor idea has been implemented through takedown notices. Copyright holders see their work posted without permission, so they send a notice to the hosting service to take it down. The service can then investigate and seek a response from the poster and determine if the work stays down or goes back up. It sounds reasonable. But if you weren't expecting a twist, you haven't been paying attention.

Large copyright holders like Disney and ViacomCBS have turned to some, by now, familiar tools to ensure their intellectual property is not given away—

namely, machine learning and AI systems that scan sites like YouTube and Facebook for their work. If they find occurrences, the software automatically generates a takedown notice. Google in 2008 received 2 million such takedown notices . . . a day.

These automated tools are not always ready for prime time. As the Electronic Frontier Foundation noted when reviewing a pending legal case against abusive copyright infringement notices:

> They describe Warner "robots" sending thousands of infringement accusations to sites like the now-closed Hotfile without human review, based primarily on filenames and metadata rather than inspection of the files' contents. They also show that Warner knew its automated searches were too broad and that its system was taking down content in which Warner had no rights—likely a violation of the DMCA.[20]

The reason automated copyright checks don't work, aside from bad programming, is that copyright is not a cut-and-dried problem. Not every use of a copyrighted work is infringement. As I said, copyright was to provide a reward and enter ideas into the public discourse. Congress wrote in specific exemptions to copyright. For example, teachers at colleges and schools can use protected works like books and articles in their classes. Satire and parody are allowed, so when shows like *The Daily Show* or *Jimmy Kimmel Live!* show clips from Fox news or the latest blockbuster, they are not breaking copyright.

Then there is the concept of fair use. Fair use recognizes the need to think and talk and incorporate ideas from protected works into everyday life. Throughout this book, for example, I include quotes from other books and articles. Not being able to do so would severely limit scholarship and public debate.

Criteria for fair use include how much of a protected work is used. A few sentences from a book is different than copying chapters. Is the new creation for commercial purposes, or will it hurt the commercial prospects of the original? Is the new work transformational, or simply a copy? And so on. But here's the rub. Breaking copyright is not like burglary or murder. Police don't show up and investigate. Copyright claims are ultimately a matter for courts and lawsuits.

All those takedown notices are more or less threats of litigation. And when a large multibillion-dollar corporation threatens you, or a website that may or may not even be operated for profit does so, it can be an act of intimidation. Copyright claims have been made to remove negative reviews of products, preserve company secrets, take down protest sites, and remove parodies that offended the person being parodied. Copyright is power: Power to make profit, and power to restrict criticism.

It is also worth noting that media companies, many of which are charged with reporting on the nation's politics and governance, are some of the largest contributors to political campaigns. In 2010 OpenSecrets.org, a not-for-profit that tracks political spending and contributions, found that:

> These organizations have—either through corporate treasuries, sponsored political action committees or both—donated almost $7 million to political action committees and so-called "527 committees" during 2009 and 2010 and nearly $38 million since the 1990 election cycle.[21]

Those organizations? News Corp, owner of Fox News Channel; General Electric, then owner of NBC, MSNBC, CNBC, and Telemundo; National Amusements, then owner of CBS; Comcast Corp.; Time Warner, then owner of CNN; and Walt Disney Co., owner of ABC.

Perhaps the obvious example of the complexities of politics and the media can be seen in the case of Michael Bloomberg. In 2020 the billionaire ran for president of the United States. He funded his campaign with his personal wealth, a wealth built up as the owner of the media company that bears his name: Bloomberg L.P. Unlike the "yellow journalism" era of the late 1800s, Bloomberg did not turn his news operations into a campaign organization. In fact, quite the opposite: the long-standing policy of Bloomberg news operation is not to cover Michael Bloomberg as a businessman, a private citizen, or a candidate to avoid potential conflicts of interest.[22]

The complex effect, however, is that this actually removed scrutiny of the candidate for president, and of his rivals. Bloomberg's campaign funding strategy also had another effect on the media—in 2019 and 2020, Bloomberg made the largest buy of television advertising in political history. The net effect was that he drove up the prices and decreased the availability of ad time on local stations around the nation for his rivals.

To be fair, copyright is the law, and every organization has a right to lobby for its own interests. As I have talked about earlier, politics and ideology are nothing new in media and, at least in the United States, the media has gone to great lengths to separate out the pressures of advertising and ideology in reporting since the days of yellow journalism. That is, until they didn't. Fox News and MSNBC are only two of the news networks that have openly ideological affiliations. And below the big media companies operate innumerable conservative and liberal news outlets, from Breitbart to the Daily Kos.

FROM CITING TO INDOCTRINATION

The ideology of these sites, however, also extends to copyright. Lewis Hyde, a writer and recipient of the MacArthur Genius award, argues that there is a predominant national narrative around the lone genius creator (my phrase). The idea is that the act of inventing Mickey Mouse was the exclusive work of Walt Disney, and therefore he deserves the credit and reward. That *Catcher in the Rye* was the exclusive result of J. D. Salinger's imagination. Hyde presents an alternative view: That Disney and Salinger were absolutely creators, and perhaps geniuses, but they built upon the "gifts" of other creators and culture. That the creative act is as much re-creation as creation. Salinger was informed by the writers before him, such as F. Scott Fitzgerald and Ernest Hemingway.[23] Disney drew upon the work of cartoonist Winsor McCay.[24] In this perspective, what does it mean to infringe copyright?

There is on old joke in the illustration field: if I copy them, they are influences; if they copy me, they're thieves. It represents the tension between Hyde's narrative of cultural gift and the narrative of creative genius popularized over a century ago by inventors like Thomas Edison. Edison built a patent-building enterprise and held 1,093 patents on everything from the electric light bulb to the printing-telegraph apparatus that was the ticker machine.

This narrative of lone genius pervades the knowledge infrastructure. Just look at the concept of plagiarism. K–12 schools and colleges are now filled with literacy instruction that goes well beyond reading to include information literacy and media literacy. These curricula seek to prepare students to be powerful participants in the knowledge infrastructure. However, all too often they instead craft students into great consumers of information and media. Instead of talking about copyright and citing work as a means of providing credit to one's influences and utilizing fair use, universities have built up an infringement regime under the banner of "academic integrity."

Wonkhe, a site that discusses higher education policy, talks about higher education and plagiarism: "From a broad historical perspective, the concept of plagiarism in universities makes no sense. Not that academic life is beyond reproach on the matter, more that the direct duplication of already existing ideas was pretty much the original point of the whole enterprise."[25]

Universities seek to teach critical thinking, impart skills, and ultimately prepare a new generation to push society forward, but a lot of that is standing on the shoulders of giants. Too much of academic integrity is about ensuring the giants are properly cited instead of critiqued, parodied, and transformed.

Don't get me wrong. Failing to give credit to influences, avoiding the hard task of learning by simply copying others without analysis, and straight-out cheating are antithetical to the academic enterprise. However, uncritical enforcement of a copyright narrative that favors owners over critical analysis is, in fact, a greater sin.

FROM CREATIVE COMMONS TO WALLED GARDENS

I realize that, unless you are reading this book in 2195, you are reading a creative act protected by copyright. As an author, I benefit from the laws that prevent wide-scale copying of my work or the selling of pirated copies on sketchy international sites. I am well aware that the benefits of intellectual property protection are what make possible the hundreds of millions of dollars budgeted to the production of the blockbuster movies I enjoy. I understand that the $9.5 billion U.S. gaming industry employs thousands of people, from digital artists and programmers to janitors and plumbers. As an overwhelmingly service-based economy, copyright and patents drive U.S. economic development. Without patents there would be no Apple, no Google, and no Disneyworld. Yet, can't we also acknowledge that without remixing and fair use, there would be no news, no hip-hop or jazz, no Andy Warhol?

Perhaps the most troubling knowledge infrastructure issue in regard to the rise of copyright conglomerates has nothing to do with copyright at all. It is the rise of closed systems for the distribution and display of media. Copyright is the default protection for creative expression, but it is not the only one. Copyright is trumped by contract law. That is, creators and consumers can sign away their rights under a contract, like a license or like the end user license you click "agree" to without even reading the text.

Narratives, entertainment, and commentary are increasingly limited to those who can pay. There was a time when music and books would be made available through safety net organizations like public libraries. Now Netflix originals, iPhone apps, and exclusive tracks on Spotify are only available in the walled gardens of content providers—beyond the reach of fair use and education not by law, but by end user licensing agreements and monthly subscription fees. The vital social transcript of culture is only available to stream for a monthly fee.

In many ways we have gone back to the days before the penny press, when only the elite could afford the subscriptions. How can we bring a balance to the knowledge infrastructure between rewarding creation and rewarding the sharing of those creations? I'll spend a lot more time on this in the "Society" section, but for now let me present part of the solution: radicalizing the stewards of the knowledge infrastructure—reporters, librarians, and professors.

NOTES

1. Brian Spear, "Submarine Telegraph Cables, Patents and Electromagnetic Field Theory," *World Patent Information* 25, no. 3 (2003): 203–9, https://doi.org/10.1016/S0172-2190(03)00072-3.

2. "Drug Patent Life: How Long Do Drug Patents Last?" Drug Patent Watch, accessed July 23, 2020, https://www.drugpatentwatch.com/blog/how-long-do-drug-patents-last.

3. Carey Dunne, "How the Television Has Evolved," Fast Company, July 22, 2014, https://www.fastcompany.com/3033336/how-the-television-has-evolved.

4. Brian Stelter, "Ownership of TV Sets Falls in U.S.," *New York Times*, May 3, 2011, https://www.nytimes.com/2011/05/03/business/media/03television.html.

5. Sean Braswell, "The Cigarette Company That Reinvented Television News," OZY, February 22, 2019, https://www.ozy.com/flashback/the-cigarette-company-that-reinvented-television-news/92498.

6. Ibid.

7. Ibid.

8. Milton J. Bates, Lawrence Lichty, Paul Miles, Ronald H. Spector, and Marilyn Young (eds), *Reporting Vietnam, Part 1: American Journalism 1959–1969* (New York: Penguin Putnam, 1998), 581–82.

9. For more on the Black press and its role in the civil rights movement, I recommend Gene Roberts and Hank Klibanoff, *The Race Beat: The Press, the Civil Rights Struggle, and the Awakening of a Nation* (Westminster, MD: Knopf Doubleday Publishing Group, 2008).

10. Ibid.

11. Chimamanda Ngozi Adichie, "The Danger of a Single Story," TEDGlobal, July 2009, 18:34, http://www.ted.com/talks/chimamanda_adichie_the_danger_of_a_single_story?language=en.

12. Carole E. Scott, "The History of the Radio Industry in the United States to 1940," EH.Net, March 26, 2008, http://eh.net/encyclopedia/the-history-of-the-radio-industry-in-the-united-states-to-1940.

13. Data provided by "Broadcast Station Totals," FCC (Federal Communications Commission), last modified November 13, 2018, https://www.fcc.gov/media/broadcast-station-totals.

14. "Number of Commercial TV Stations in the United States from 1950 to 2017," Statista, December 2017, https://www.statista.com/statistics/189655/number-of-commercial-television-stations-in-the-us-since-1950.

15. Katerina Eva Matsa, "Buying Spree Brings More Local TV Stations to Fewer Big Companies," Pew Research Center, May 11, 2017, https://www.pewresearch.org/fact-tank/2017/05/11/buying-spree-brings-more-local-tv-stations-to-fewer-big-companies.

16. Ibid.

17. Some examples: Dylan Matthews, "Sinclair, the Pro-Trump, Conservative Company Taking Over Local News, Explained," Vox, April 3, 2018, https://www.vox.com/2018/4/3/17180020/sinclair-broadcast-group-conservative-trump-david-smith-local-news-tv-affiliate; Sheelah Kolhatkar, "The Growth of Sinclair's Conservative Media Empire," *New Yorker,* October 22, 2018, https://www.newyorker.com/magazine/2018/10/22/the-growth-of-sinclairs-conservative-media-empire; Eric Levitz, "Pro-Trump Sinclair Media Poised for National Expansion by 2020," Intelligencer, April 23, 2019, https://nymag.com/intelligencer/2019/04/sinclair-broadcast-group-national-expansion-2020-sports-networks.html; and Eli Rosenberg, "What We Know about the Conservative Media Giant Sinclair," *Chicago Tribune,* April 30, 2018, https://www.chicagotribune.com/business/ct-biz-who-is-sinclair-broadcast-group-20180403-story.html.

18. Matthews, "Sinclair, the Pro-Trump, Conservative Company Taking Over Local News, Explained."

19. Ashley Lutz, "These 6 Corporations Control 90% of the Media in America," *Business Insider*, June 14, 2012, https://www.businessinsider.com/these-6-corporations-control-90-of-the-media-in-america-2012-6.

20. Mitch Stoltz, "In Hotfile Docs, Warner Hid References to 'Robots' and Its Deliberate Abuse of Takedowns," Electronic Frontier Foundation, October 9, 2014, https://www.eff.org/deeplinks/2014/10/hotfile-docs-warner-hid-references-robots-and-its-deliberate-abuse-takedowns.

21. Megan R. Wilson, "Not Just News Corp.: Media Companies Have Long Made Political Donations," OpenSecrets.org, August 23, 2010, https://www.opensecrets.org/news/2010/08/news-corps-million-dollar-donation.

22. Bob Garfield, "Bloomberg on Bloomberg," February 21, 2020, in *On the Media*, produced by WNYC Studios, podcast, MP3 audio, 15:23, https://www.wnycstudios.org/podcasts/otm/segments/bloomberg-bloomberg.

23. Kathy Gabriel, "The Influence of Ernest Hemingway and F. Scott Fitzgerald on J. D. Salinger," Salinger in Context, December 4, 2010, http://salingerincontext.org/category/salinger-in-context/influences.

24. Jeff Suess, "Enquirer Artist Influenced Walt Disney," *Enquirer*, March 19, 2014, https://www.cincinnati.com/story/news/2014/03/19/enquirer-artist-influenced-walt-disney/6608049.

25. David Kernohan, "A Brief History of Plagiarism and Technology," Wonkhe, March 20, 2019, https://wonkhe.com/blogs/a-history-of-plagiarism-and-technology.

10

Trust

From Walter Cronkite to Saul Alinsky

The mission of the CS *Alert* was not accompanied by a pool of reporters or a camera crew. There were no photographs, and the only recognition the *Alert* and her crew received during the war was from the cheers of the French destroyer group on the morning of August 15.

That is not to say there wasn't support for the war. Many throughout Europe supported not only the war effort, but the idea of war in general. That may seem counterintuitive considering the toll the Great War took in lives and misery. Today we see images of total war, with conscripted soldiers fighting for inches in trenches near destroyed cities. However, at the time the war was supposed to last for only months and well away from populated places.

Up to World War I, wars in Europe tended to be much different. They were smaller, contained to rural areas, and fought by professional soldiers. The previous European war, the Franco-Prussian War in 1870, lasted a little over six months. It was, however, hardly bloodless, with approximately 250,000 dead—about 162,000 from a smallpox epidemic spread by French prisoners of war.[1]

Yet one has to be careful when using the phrase "support for the war." In today's knowledge infrastructure we are used to something quite different when presenting evidence of support or when describing public sentiment. For example, following 9/11 an astounding 92 percent of Americans supported military action against the attackers, including going to war; 84 percent supported a broader military campaign against any country that assisted or supported terrorism.[2] On September 22, 2001, President George W. Bush enjoyed an approval rating of 90 percent, the highest presidential approval rating ever recorded.[3]

In 1914 there were no such polls or data available. Support was determined by a very unscientific process of analyzing newspaper opinion columns, the letters of prominent individuals, statements by politicians, votes in government and, of course, the fact that these countries did go to war. The modern-day concept of opinion polling simply wasn't the norm at the time. How we have come to rely on polling data, and what such data tells us about the perceptions of knowledge workers, is the focus of this chapter. To start that journey, we have to step back in time beyond 1914 to 1824 and the presidential race of Andrew Jackson versus John Quincy Adams.

FROM STRAW POLLS TO MARGINS OF ERROR

In 1824 the *Harrisburg Pennsylvanian* newspaper conducted the first political poll in the United States. It accurately predicted the winner of the popular vote, Andrew Jackson. Of course, as we have become accustomed to, the popular vote doesn't guarantee the presidency, and John Quincy Adamas was selected as president by the House of Representatives.[4]

Even though this poll presented data in a similar fashion to what we are used to today, these results were not the same. The type of political polling conducted by the *Pennsylvanian* is known as a straw poll. It was a simple count of people available to give their opinion—what in science we would call a convenient sample (as in, who was convenient to ask). Another early polling effort that was a clear step up in terms of time and effort were the straw polls of the *Literary Digest*. These polls of voters around politics and elections were the primary means of gauging political sentiment from 1916 to 1936. At their height, the *Digest* sent out over 20 million "ballots" based on names from telephone directories and automobile registrations. The polls were a regular fixture of newspapers at the time. It would take over a century to pass from the *Pennsylvanian* poll to the birth of modern polling we are accustomed to today.[5]

Much of modern opinion polling can be attributed to George Gallup. Gallup had a PhD in mathematics and saw the potential for polling in journalism and advertising. He began his career as a professor of journalism, but in 1932 he moved to New York City and joined the advertising agency Young & Rubicam as director of research. Gallup put his considerable skill in mathematics and statistics to the selling his method of determining public opinion to advertisers. His work injected a sense of public reality to the campaigns of the time.

In 1935, Gallup formed the American Institute of Public Opinion. He cemented both his role and the role of public polling in elections in 1936 when he correctly predicted the election of Franklin Delano Roosevelt against the *Literary Digest* poll that predicted Alf Landon would win.

The fundamental difference between Gallup's polling and the long-standing straw poll was called quota sampling—what we today would call scientific sampling. Rather than just ask a lot of people a question, the goal was to ask a smaller, but more representative sample. The 20 million ballots of the *Literary Digest* failed to predict FDR's win because it overwhelmingly asked the opinions of more well-to-do folks—people who could afford a car and a telephone. FDR won (by a landslide) because while he captured some of this population, a majority of poor and working-class Americans devastated by the Great Depression favored FDR's progressive policies.

Scientific polling adopted the methods of social science pioneered by scholars like W. E. B. Du Bois. Du Bois grounded his work by studying and documenting the African American experience in such works as *The Philadelphia Negro* and *The Gift of Black Folk: The Negroes in the Making of America*. These books demonstrated how the rising mass media didn't represent some universal objective reality, but a narrative crafted by a majority view—a point we will return to momentarily. Du Bois also pioneered the use of surveys, focus groups, and interviews as a scientific method for determining the views of populations. These advances were adopted into Gallup's method of survey, making the scientific method cheaper and more accurate than the alternatives.

The success of Gallup and his contemporary pollsters like Archibald Crossley and Elmo Roper led to increased sophistication in scientific and statistical polling methodologies. This, in turn, led to the increased incorporation of pollsters into political campaigns. Polls also become important to the functioning of the federal government itself. The Division of Program Surveys in the Bureau of Agricultural Economics was formed in 1922 to gather farmers' feedback on the work of the Department of Agriculture.[6] During World War

II the Office of War Information created the Division of Morale of the U.S. Army. This division did regular and extensive polling on American sentiment toward the war and a wide range of issues.

However, even with all the advances, polls were hardly as infallible as they were presented. The U.S. presidential election of 1948 still serves as an important reminder of the fallibility of polling and data in general. All the major polls (Crossley, Gallup, and Roper) projected that Thomas Dewey would beat incumbent Harry Truman for the presidency. The belief in polling was so strong, many large newspapers ran the headline "Dewey Defeats Truman," having to print and distribute their papers before the election results were in. This led to the now-famous image of a victorious Truman holding up a copy of the *Chicago Daily Tribune* on the morning of his re-election (figure 10.1).

FIGURE 10.1
Harry Truman shows off a newspaper headline that wrongly predicted the outcome of the presidential race based on polling. *https://www.flickr.com/photos/scriptingnews/2544447858*

The limitations of polls, particularly in presidential elections, were once again on full display with the election of Donald Trump in 2016, when most polls predicted the winner would be Hillary Clinton (and again in 2020 with the prediction of a "blue wave" that once again vastly underestimated the support for Trump). The soul-searching on how so many polls so wrongly predicted the outcomes is a case study in the limitations of dataism. The data gathered was accurate; it was the underlying assumptions of the pollsters that led to the false conclusion. As the Pew Research Center wrote in its initial review of the inability of polling to predict the race's outcome:

> One likely culprit is what pollsters refer to as nonresponse bias. This occurs when certain kinds of people systematically do not respond to surveys despite equal opportunity outreach to all parts of the electorate. We know that some groups—including the less educated voters who were a key demographic for Trump on Election Day—are consistently hard for pollsters to reach. It is possible that the frustration and anti-institutional feelings that drove the Trump campaign may also have aligned with an unwillingness to respond to polls. The result would be a strongly pro-Trump segment of the population that simply did not show up in the polls in proportion to their actual share of the population.[7]

Mona Chalabi of the *Guardian* was a bit more blunt in her assessment:

> The polls were wrong. And because we are obsessed with predicting opinions rather than listening to them, we didn't see it coming. So, the world woke up believing that Republican candidate Donald Trump had a 15% chance of winning based on polling predictions—roughly the same chance of rolling a total of six if you have two dice. Despite those odds, the next US president will be Donald Trump.[8]

To be clear, the scientific methods of polling that Pew and other pollsters use have a great advantage over the machine-learning, data-centered approaches I discussed in chapter 5. The method allows scientists and researchers to audit the mistakes and hopefully account for them in future work. However, polling shows in the most precise way that collected data are not neutral assets to simply be run through algorithms. Whom the data comes from matters. The assumptions of the people processing that data matter. Human agency is central—it was central in 2016 when people decided to answer a poll, and it was central when W. E. B. Du Bois examined the roles of Black Americans in reconstruction.

FROM ELECTIONS TO EDELMAN

The history of scientific polling and survey work that started out in advertising grew to encompass politics, and now covers virtually every part of our society—from economic indicators of consumer confidence, to product design and consumer satisfaction, to college strategic planning, to issues of trust. Polling companies such as Gallup and the Pew Research Center track citizen trust in different institutions and professions including the military, the media, librarians, nurses, and politicians.

Polling data, while not perfect, does provide a useful yardstick to measure the effects of data and media on society. Take, for example, the perception of conflict and the military. Unlike World War I, we now have scientific opinion polling on how the U.S. population viewed war and soldiers. Polls showed how public opinion turned in the Vietnam War in the 1960s and 1970s:

> After an initial "rally-round-the-flag" (or "round-the-President") period the American people slowly began to move away from simple acceptance of administration policy. Support for presidential policy in Vietnam declined steadily during 1966, while preferences for escalation grew. . . . In 1967, escalation became the preferred policy of most Americans for a short time. Then, in 1968, the proportion of the public which preferred this course began to drop off. Escalation continued to lose favor until 1970, when pollsters stopped asking respondents whether they favored that course. By the end of direct American involvement in Vietnam most citizens considered the prospect of renewed military efforts there as unpalatable; at the time of the Paris peace accords in 1973, 79 percent of the public opposed the reintervention of American military troops in Vietnam even "if North Vietnam were to try to take over South Vietnam."[9]

The graphs accompanying this article (see figure 10.2) show the rather stark picture of how support for the war disappeared.

This view of the war also had a direct effect on how people perceived the military. Gallup has regularly polled Americans on their trust in key institutions and professions. In 1972 the organization began to ask folks to "tell me how

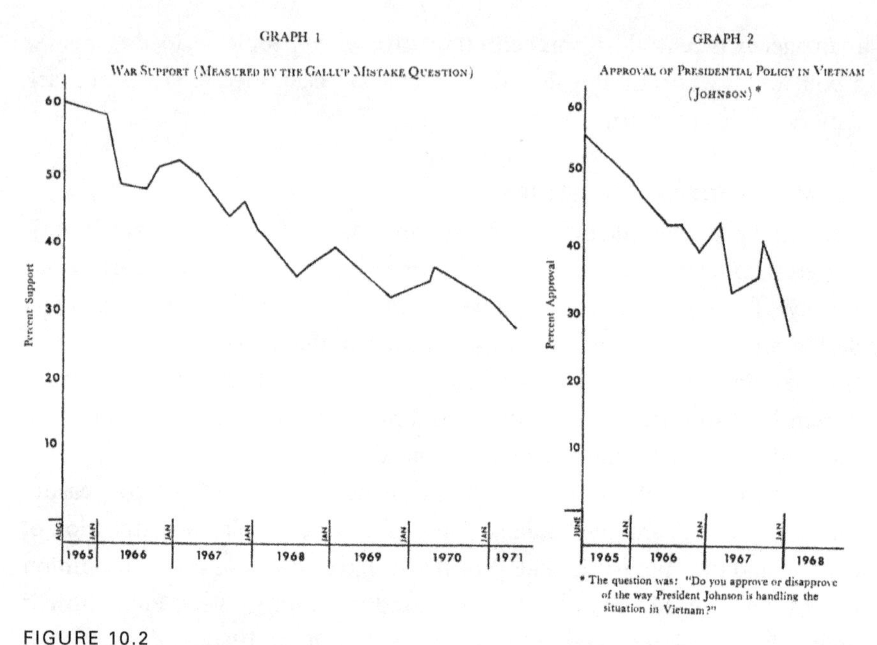

FIGURE 10.2
These graphs show a rapid decline in support for the Vietnam war and the president. *William Lunch and Peter Sperlich, "American Public Opinion and the War in Vietnam,"* Western Political Quarterly *32, no. 1 (1979): 21–44, http://doi.org/10.2307/447561*

much confidence you, yourself, have in [the military]—a great deal, quite a lot, some or very little?" In 1975 those numbers were 27 percent a great deal, 31 percent quite a lot, 25 percent some, and 11 percent very little.[10] But by 2019 those numbers had grown to: 45 percent a great deal, 28 percent quite a lot, 18 percent some, and only 8 percent very little. Graphing Gallup's results over the past 44 years (figure 10.3), we can see a steady increase in citizen trust in the military.

In fact, in 2019 the military, according to Gallup, was the most trusted institution in America, beating out television news, schools, Congress, and churches. Note in particular the spike in 1991, during Operation Desert Storm, and again from the beginning of 2001 to the beginning of 2002 (9/11). Also of interest is the spike in 2009 with the election of Barack Obama and his message of moving away from the war policies of President George W. Bush (followed by continued engagement in Iraq and a corresponding drop in military support).

This brings us right back to the knowledge infrastructure: the policies, sources, technologies, and people used to find meaning and gain power

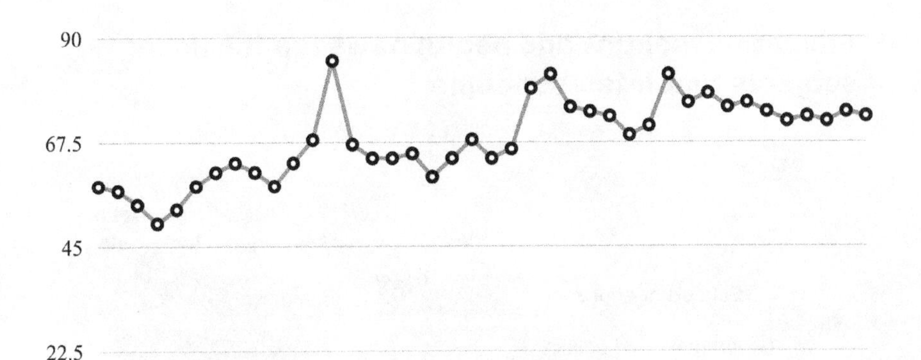

FIGURE 10.3
Trust in the military, as found by Gallup. *Original by author*

through learning. In chapter 9 we talked about some major policies that shape the infrastructure, such as copyright and patents. Most of the "Data" section of this book explored technologies such as massive-scale computing and ubiquitous networks at work in our modern infrastructure. The discussion of mass media and of quantifying the world with cheap connected microchips introduced the idea of sources in the knowledge infrastructure. Let me now turn my attention to people.

At the broadest level, everyone you know is a part of the knowledge infrastructure. Study after study shows that friends and family are big influencers of your worldview on topics including politics, education, even the level of optimism you bring to your daily activities. But for this discussion, let me single out groups of people that actively and knowingly curate the knowledge infrastructure. That is, professionals who seek to inform and educate. Yes, people like reporters, librarians, and teachers, but also people like doctors and nurses (medical information) and politicians and civil servants.

The Pew Research Center has been investigating the issue of trust and information for over a decade. The center uses scientific polling techniques to see how people are building their knowledge and the sources they use to do it. For example, in 2017 Pew found that education topped the list of subjects American citizens were interested in, followed by government, health, and technology (see figure 10.4).

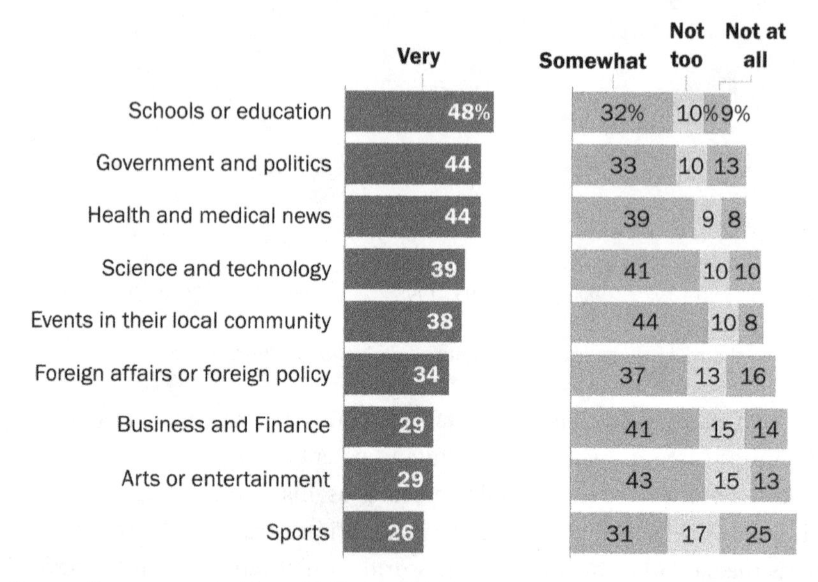

FIGURE 10.4
Pew Research Center results on subjects of interest to U.S. adults. *https://www.pewresearch.org/internet/2017/09/11/the-elements-of-the-information-engagement-typology/*

Pew then asked about how much trust those polled had in key sources of information. The results were concerning. Librarians topped the list, followed by health care providers and then family and friends. Good news for these folks, except that the top scorer—librarians—only had "a lot" of trust from 40 percent of respondents. Reporters, by the way? Fourth for local news, and sixth for national news. At the bottom? Social media sites (see figure 10.5).[11]

Gallup also regularly conducts surveys on the public perceptions of different professions. Asking a slightly different question—"How would you rate the honesty and ethical standards of people in these different fields"—Gallup found that nurses topped the list (85 percent), followed by engineers (66 per-

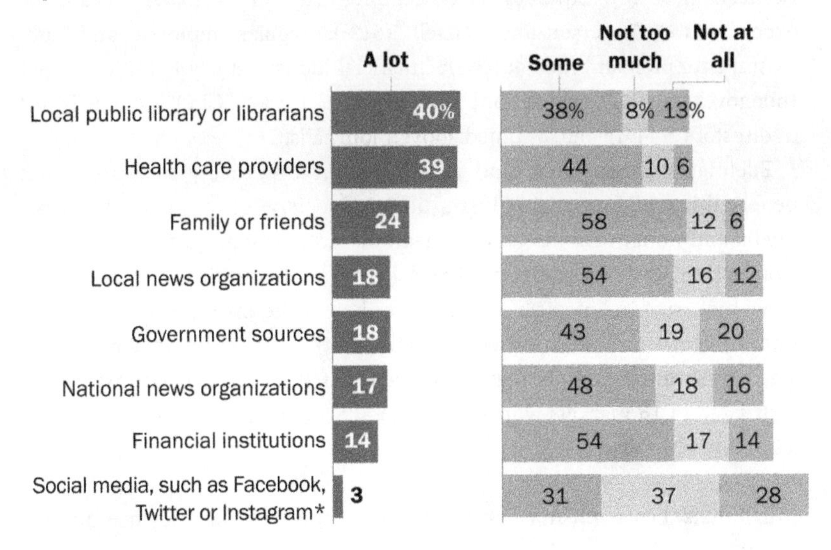

FIGURE 10.5
Pew Research Center results on U.S. adult trust in select information sources. *https://www.pewresearch.org/internet/2017/09/11/the-elements-of-the-information-engagement-typology/*

cent), medical doctors (65 percent), and pharmacists (64 percent). Professors came in at 49 percent; journalists, 28 percent. Members of Congress, 12 percent and "Car salespeople" at 9 percent.[12]

The News Literacy Project reported on a poll by Edelman, a communications company, on trust in professions and institutions globally. Reporters did not fare much better in this poll. Of particular note in this study was the separation of journalists (people) from media (institutions); the high ranking of "a person like yourself"; and the definition of the media as "the institution focusing on our 'information and knowledge' well-being." Professionals focusing on our information and knowledge well-being is not a bad definition of people in the knowledge infrastructure.

The most trustworthy source for information, according to 65% of the respondents, was a "company technical expert." "Government official" was deemed least credible, with barely a third (35%) trusting someone in that position. Between those two extremes, in descending order of trust, were "academic expert" (63%); "a person like yourself" (61%); "regular employee" and "successful entrepreneur" (tied at 53%); "financial industry analyst" (52%); "NGO [nongovernmental organization] representative" (48%); "CEO" (47%); "board of directors" (44%); and, as noted above, "journalist."

Edelman's annual trust and credibility survey broadly measures what people think of four major institutions in society: government, business, media and nongovernmental organizations. Interestingly, although trust in "journalists" is down globally, when asked about institutions, the responses show that trust in "media" (loosely described as the institution focusing on our "information and knowledge" well-being) is up from the year before. The not-so-great news is that media remains the least-trusted institution, distrusted in 16 markets worldwide. The U.S. is one of those 16 markets, at a level of 48% "distrust."[13]

Journalists, however, may find comfort in the fact that they are not alone; since 1975, Gallup found marked declining trust in newspapers, TV news, public schools, churches, and the president (see figure 10.6).

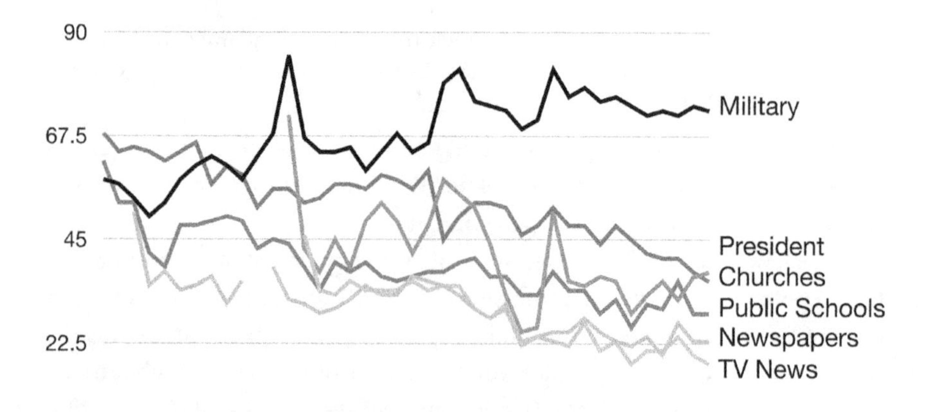

FIGURE 10.6
Trust in the military has risen as trust for the U.S. president, churches, public schools, newspapers, and TV news have declined. *Original by author*

You may think this is a result of people just being less trusting. However, Edelman found a global increase in trust among the general population and those they classify as "informed public": those people ages 25–64 with a college education, in the top 25 percent of household income in their countries, and report "significant media consumption and engagement public policy and business news." In the United States, the overall trust index was 49 (out of a possible 100), up 6 points in the general population from 2018 to 2019; the index was up 15 points among the informed public, to 60.[14]

So why do people seem to trust the military, nurses, and librarians over journalists, the government, and car salesmen? And what is the implication of this for the knowledge infrastructure? It comes down to the nature of knowledge and trust.

FROM CHANGING THE CHANNEL TO THANK YOU FOR YOUR SERVICE

The answer to why some professionals are trusted and some are not is complex. Trust, for example, has notable neurological components. As infants we are literally wired to bond with our mothers and seek out social connection.[15] There is evidence that controlling the level of the chemical oxytocin in the body influences how trusting a person is. Why trust varies from profession to profession and country to country, however, is impossible to concretely state. There are too many variables. However, we have seen in this section of the book that how narratives are developed in the media, how they are targeted toward certain people, and how those narratives are connected to worldviews play a strong role.

Teachers and Russian trolls alike use the psychological concept of *confirmation bias*. Confirmation bias is the confirmation of and support for information that fits with a person's existing view of a situation. Chat bots and foreign election interference alike often begin their work with spreading factual news that would appeal to their target audience. Once friended or followed, then dissemination of misinformation begins.

We also know that trust has a strong social component, and that inclusion in groups leads to greater trust. We trust people whom we find similar to us. On a grand scale. this means things like race, gender, socioeconomic status, national origin, and the like. However, it operates on a very basic level:

> We're far more likely, for example, to trust people who are similar to us in some dimension. Perhaps the most compelling evidence of this comes from a study

> by researcher Lisa DeBruine. She developed a clever technique for creating an image of another person that could be morphed to look more and more (or less and less) like a study participant's face. The greater the similarity, DeBruine found, the more the participant trusted the person in the image. This tendency to trust people who resemble us may be rooted in the possibility that such people might be related to us. Other studies have shown that we like and trust people who are members of our own social group more than we like outsiders or strangers. This in-group effect is so powerful that even random assignment into small groups is sufficient to create a sense of solidarity.[16]

Lee McIntyre, interviewed on the *On Point* radio show,[17] tells the story of Jim Bridenstine, who as a congressman from Oklahoma openly called climate change a hoax. President Trump appointed Bridenstine as NASA administrator. A few months into his tenure at NASA, Bridenstine acknowledged that climate change was a reality and confirmed his confidence in the scientific evidence. While this certainly could be an attempt to quiet critics, it is more likely that his time working and getting to know NASA personnel and scientists had a great deal to do with his reversal.

The reality of trust in the knowledge infrastructure matches the idea that, as people, we are driven by a need to learn. The same factors that lead to trust—connecting to what we already believe, connecting to shared narratives, sources of information that match our identity, and trust as constructed—are the exact same factors needed for learning. We learn by scaffolding—connecting something new to something we already know. We learn through narratives. We learn from systems where we feel we belong. But most importantly, we learn in systems where we feel ownership.

Abraham Maslow was a psychologist who studied motivation—particularly what motivated people to be curious and become satisfied in themselves. His "Hierarchy of Needs" (figure 10.7) has become fundamental in education psychology and learning. At the bottom of his hierarchy, presented as a pyramid, are "physiological" needs. People won't learn (or trust) if they are hungry. In order to become curious, basic needs must be met. Next is "safety," which includes personal security, emotional security, financial security, and health. Next is "belonging and love," which includes friendship and family. "Esteem," is the feeling of recognition and pride. At the top is "self-actualization," which is the realization of one's full potential.

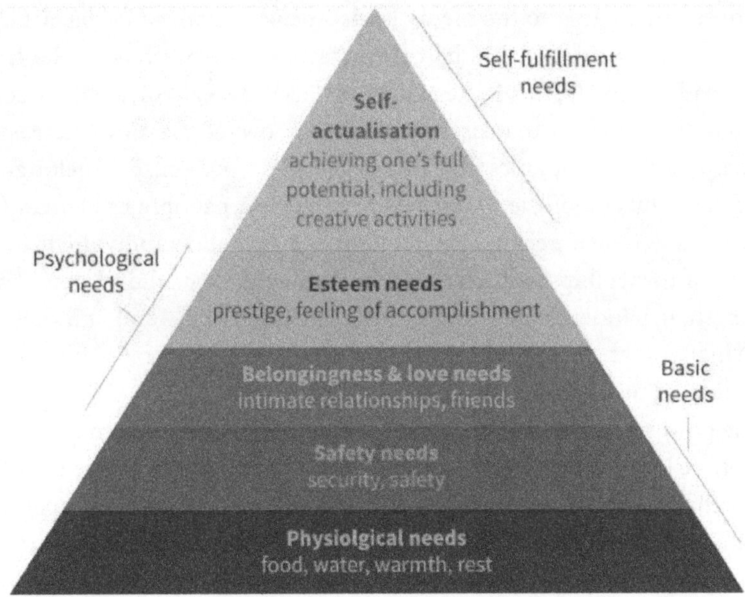

FIGURE 10.7
Abraham Maslow's hierarchy of needs as visualized by Chiquo. *https://en.wikipedia.org/wiki/Maslow%27s_hierarchy_of_needs#/media/File:Maslow's_Hierarchy_of_Needs.jpg*

Take a look, for the moment, at the "belonging and love" needs. These fit with what we have seen throughout this chapter. It also begins to explain the U-shaped curve in science learning discussed in chapter 6, where a part of required education in primary and secondary school learning fails. Yet when it is part of our own interest or job (keeping said job is very much in our own interest), learning goes up. We want to feel ownership and membership.

Now think about the difference between trust in the military and trust in media. In the United States, the military has evolved since the Vietnam War. The forced belonging of the draft has been replaced with volunteers and career soldiers. Through concerted efforts of messaging, marketing, and propaganda during the Gulf Wars, the military is seen as an important part of being American. The military now pays, for example, to be a part of football games and sporting events. That flag ceremony before the Super Bowl? It actually costs the taxpayers to honor the flag. Also, there is only one military, with little direct choice in its leadership or organization. Our choice is simple: support this one or, well, don't.

Now compare that to the media landscape we discussed in the last chapter. Media consolidation has had two effects on media choice. The first is that broadcast media has increasingly replaced a local voice with a corporate one. Newspapers that used to be locally owned are now increasingly owned by a few companies and individuals.[18] So our sense of belonging is decreased. Outside of the major news outlets is a panoply of choices from blogs to Reddit channels to Facebook groups that allow individuals to find a sense of ownership and belonging in narrow streams. Local news scores higher than national news. Media that includes blogs and internet sites score higher than trust in reporters who are associated with the increasingly consolidated media corporations.

The power of ownership plays out in churches as well—membership is at an all-time low.[19] In schools we see an increase in parent involvement, but met with a charter schools movement that has in places successfully propagated a narrative of school choice and the dangers of increased regulation in public school governance. We could also look at the declining trust in the presidency, Congress, and the Supreme Court as a sense of disconnection from federal politics and increased isolation into ideologies.

All of this is met with a growing sense of consumerism. Relationships are increasingly seen as transactions. We pay for our schools, we consume our news, we exchange our vote and donations for access. Yet we join the military. We thank soldiers for their service and welcome them home to families.

We see a recognition of this need for ownership, belonging, and the connection to learning in the growing use and reputation of public libraries. Libraries globally are experiencing a sort of renaissance. Library usage in the United States and around the world is increasing.[20] The field of librarianship represents an annual investment of nearly $26 billion in North America and well over $40 billion worldwide. In an age when traditional institutions are declining, library usage has grown steadily over the past 20 years. "One out of every six people in the world is a registered library user" and "five times more people visit U.S. public libraries each year than attend U.S. professional and college football, basketball, baseball and hockey games combined."[21] Americans go to school, public, and academic libraries more than three times more often than they go to the movies. Almost 100 million more people visit their libraries (1.3 billion visits) each year than see a movie at the theater (1.2 billion views).[22]

In Europe, cities such as Oslo, Utrecht, Tilburg, Aarhus, and Manchester have built grand new central libraries. A recent study in Norway found that as governments are pushing more and more services online, the use and need for physical libraries is increasing.[23]

The growth of institutions once thought headed toward obsolescence has also drawn the attention of scholars and the media outside of librarianship. Eric Klinenberg, a sociologist at New York University, calls them "palaces for the people" and a piece of vital social infrastructure.[24] Articles in major publication outlets such as the *New York Times*, the *Guardian*, CNN, and *Forbes* make it clear that libraries are not only surviving in today's connected and data driven society, they are thriving.[25] Why?

The short answer is a transformation toward the communities they serve, and a marked departure from neutrality.

Public libraries are just that—public. They are locally controlled, with the majority of their funding coming from the local community and overseen either by citizen boards or local government. Over the past decade libraries have also transformed from information centers filled with books to community hubs providing equitable access to services like literacy support, makerspaces, technology training, meeting spaces, business support, and community conversations on important topics like race, immigration, and LGBTQ rights. In gentrified neighborhoods, libraries serve as community living rooms. In rural towns, they provide high-speed internet access and access to databases and books that individuals couldn't afford on their own. As more and more government services have retreated from the public sphere or to online FAQs, librarians—real human faces—stand ready to work with community members regardless of race, religion, or the ability to pay.

Just as important in their roles in the knowledge infrastructure, however, is that these human faces and equitable service are no longer driven by a false myth of neutrality. For well over a century, librarians have seen neutrality and unbiased service as a professional principle. Recently, however, this myth has simply imploded. What was seen as neutrality was actually adherence and continuation of a predominant national narrative. Librarians could claim to be neutral because they were in line with a majority white view of what a community needed.

The collection was built in an unbiased way—except that libraries were buying books from publishers that overwhelmingly supported white authors.

Libraries were serving all who walked into their doors equally—even though too many city residents of color had neither the transportation access nor time after working multiple jobs to actually make it to the library. Story times were offered during the daytime for all—even though, according to the Population Reference Bureau, "only 7 percent of all U.S. households consisted of married couples with children in which only one spouse worked. Dual-income families with children made up more than two times as many households."[26] So, these public libraries were in effect offering a service to an elite few who could afford to either stay home with their children or have a nanny who could take the kids to story hour.

Over the decades, librarians have shifted their services and their concepts. As more material became available online, they could spend more time and space in learning. The new Oslo Central Library in Norway, for example, will have half of the physical books of the previous building. That makes space for community collaboration areas (theaters, meeting rooms, open performance spaces). In Finland and Norway, new national legislation directs public libraries to host community-wide conversations on democracy and politics. The library is the new town square.

We are told that the loss of trust in institutions and professions is due to rejection of truth. Many see the answer to a fragmented public in terms of news and politics as a return to objectivity and facts. What librarians have discovered is that the way to trust is through being transparent in their biases and being advocates for their communities.

Public libraries are not neutral. Librarians believe in providing the widest possible access to the widest possible set of sources. That is not neutral; that is an anti-censorship stance. Librarians believe in the importance of privacy. As I talked about in relation to the Patriot Act, that is not a neutral stance. Librarians build relationships with communities, determine needs and aspirations, and then work with the community to address them. Librarians are increasingly throwing off transactions (how many books loaned out, how many questions asked) for relationships (how many partnerships formed).

Take the unrest in Ferguson, Missouri, in 2014. The shooting of an unarmed Black teen, Michael Brown, by a white police officer, Darren Wilson, led to mass protests. A militarized police force—clad in body armor, helmets, and camouflage—shot rubber bullets and tear gas at protestors. Children huddled in their homes, unable to sleep as their parents took turns watching the front door for trouble, one father sitting next to a baseball bat "just in case." The Missouri governor called up the state police and National Guard, announced curfews, and closed government institutions.

This was the majority of reporting in Ferguson. It is disturbing to be sure—disturbing because this was not happening overseas, but in the suburbs of the U.S. heartland. Issues of justice, race, and economic disadvantage had jumped from an unspoken "issue" to the front of America's consciousness. Yet in the images of protestors doused in tear gas and armored police transports another story emerged: with the closure of most public institutions in Ferguson, the children of the town were out of school.

This was not just a matter of a delayed school year; for many low-income families, this was a matter of food. A large percentage of Ferguson's youth receive food assistance through the schools. With the schools closed, these children went hungry—hungry and trapped in their homes with the sounds of gunfire and exploding tear gas canisters outside.

The Florissant Valley Branch of the Saint Louis County Library and the local Ferguson Public Library stepped up to help. Both libraries share coverage of the Ferguson-Florissant School District. Both showed bravery, and both showed that librarians can be change agents.

Jennifer Ilardi came into her library on Tuesday, August 19, and decided to set up a variety of art supplies in the library's auditorium so that parents could have some activities, get out of the house, socialize, and create. She also decided to order pizza. During a TV interview, she was prompted with, "So you saw a need in the community. You saw a void." She responded, "This is what libraries do. We supplement our educational system regularly with after school programs and summer programs. We provided free lunches all summer long through a collaboration with Operation Food Search because we recognize that a large portion of our community qualifies for free and reduced price lunches."[27]

The librarians continued programs until the schools of Ferguson were reopened. Operation Food Search agreed to continue the lunch program. The Magic House, a local children's museum, offered free interactive educational activities for students. Local artists volunteered their services as well, offering free magic shows. Scott Bonner, director of the Ferguson Public Library down the street, teamed with the teachers of the closed school to hold classes in the library. When they ran out of room, the librarians connected with local churches and youth centers to make a new, impromptu school. None of these acts were neutral. All of these acts were local.

FROM CLASS RINGS TO CLASS RATINGS

Compare this long-term transformation occurring in public libraries to another set of institutions integral to the knowledge infrastructure: higher education. As we discussed in chapter 6, colleges and universities—particularly research universities—have been transformed by federal research dollars, a process that began in World War II. The introduction of research funding, decreased state funding, new competition in the form of for-profit universities, and a particular kind of "cost disease" are pushing universities to be more and more oriented toward transactions and justifying the high price of college. This shift is having an effect on trust in colleges and universities.

The disease I refer to is Baumol's cost disease. William Baumol was an economist, and he noticed a particularly interesting exception to labor productivity. In economics, productivity is the amount of product or service that can be delivered by a worker. It is so central to the working of our economy it is often reported without explanation. Technology, economies of scale, and the amount of competition in a market all drive different industries to produce more with less. It is why products cost less, but companies can increase profits over time.

Baumol looked to apply ideas of productivity to the service industry, which is based on people power, not manufactured goods. The "disease" he found was that in many industries it took the same amount of effort (and people) to produce the same effect, no matter the technological advances or efficiencies. For example, it takes the same number of orchestra members the same amount of time to play Beethoven's Fourth Symphony as it did in Beethoven's time. Yet the cost of the members of the orchestra (their wages and benefits) have gone up considerably. In higher education, a professor can't teach the same number of students faster this year than last. The net effect is that these service-based industries will increase in cost over time. And in fact, if you look at the cost of service industries (figure 10.8), that's exactly what happens.

While some have disputed the scale of this effect in higher education (and the idea that teaching can't be more efficient), the fact remains that a college education is getting more expensive. A lot more.

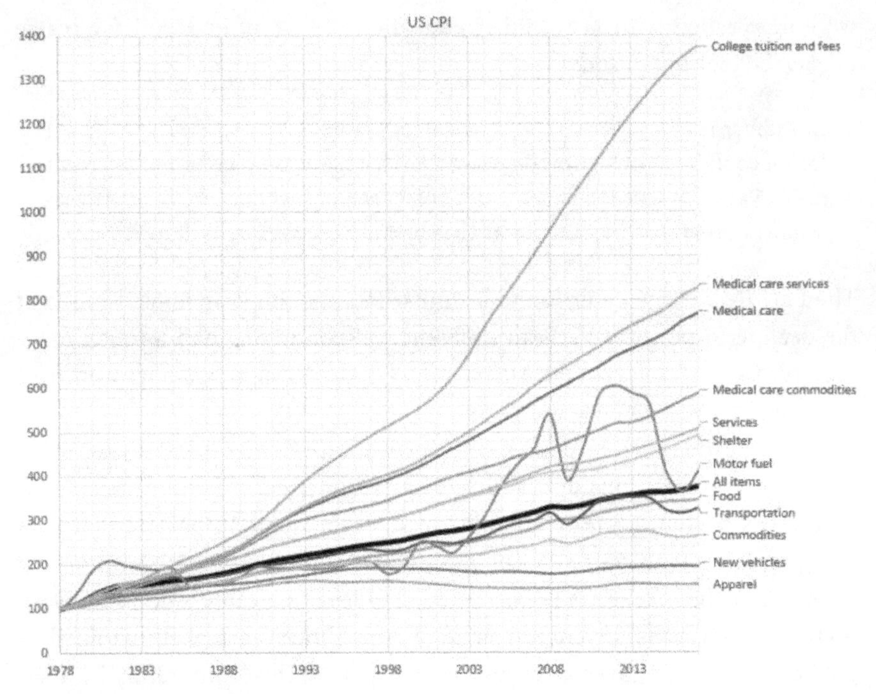

FIGURE 10.8
Increases in costs for services versus costs for manufacturing. *https://commons.wikimedia.org/w/index.php?curid=75468943*

The problem of cost in higher education is compounded by the fact that a research university is not just about teaching. Costs in universities also come from research and service to the community. These factors are often not taken into consideration by the statehouses that fund universities, nor by students shopping for a degree.

The net result is that higher education becomes more and more expensive, while more and more stakeholders are seeking to define the value of college more and more narrowly. The value of colleges is increasingly reduced to the jobs and salaries students can get when they graduate. This is not all bad (though later I will argue it is too narrow), but it sets up education as a transaction, not a relationship.

We are beginning to see the negative effects in terms of trust showing up in views of higher education. Students are increasingly seen and see themselves as consumers of college.[28] In a recent poll, 65 percent of respondents were unsatisfied with U.S. higher education.[29] Gallup, in a poll for *Inside Higher Education*, found:

> Just under half (48 percent) of American adults have "a great deal" or "quite a lot" of confidence in higher education. . . . That figure is down from 57 percent in 2015 and represents a larger than typical decline in confidence in an American institution in a relatively short time period, according to Gallup.[30]

The Pew Research Center found a similar fall, though it attributed much of the decline to political partisan differences; Republicans had a much more negative view of colleges than Democrats, who had a slightly higher view over the past eight years.[31]

In the reverse of the public library case, universities seem to be moving away from relationships, and so away from trust. Whereas the view of libraries is becoming more complex—the nostalgic book palaces narrative being replaced with narratives of vital social infrastructure and community hubs—views of college are being flattened to teaching centers geared toward preparing graduates for employment. Where libraries are well ingrained in public finance and governance and seen as part of the "community," higher education is increasingly seen as a product often funded by research dollars, donations, and sports programs. Therefore, when librarians shed the idea of neutrality, they replace it with trust based on relationships, but when colleges and the media do so, they are seen as self-interested and biased—as flawed products in a sea of other options.

As we continue our discussion of the knowledge infrastructure, the ideas of neutrality by those who seek to shape and maintain it will be central. The idea of relationships over transactions will be central. People will be central.

NOTES

1. Micheal Clodfelter, *Warfare and Armed Conflicts: A Statistical Encyclopedia of Casualty and Other Figures, 1492–2015*, 4th ed. (Jefferson, NC: McFarland, 2017).

2. Eric V. Larson and Bogdan Savych, "American Public Support for U.S. Military Operations from Mogadishu to Baghdad," RAND Arroyo Center, 2005, https://www.rand.org/content/dam/rand/pubs/monographs/2005/RAND_MG231.pdf.

3. Marc J. Hetherington and Michael Nelson, "Anatomy of a Rally Effect: George W. Bush and the War on Terrorism," *PS: Political Science and Politics* 36, no. 1 (2003): 37–42, https://www.jstor.org/stable/3649343?seq=1#metadata_info_tab_contents.

4. G. Terry Madonna and Michael Young, "The First Political Poll," *Politically Uncorrected*, June 18, 2002, Franklin & Marshall College, https://www.fandm.edu/uploads/files/271296109834777015-the-first-political-poll-6-18-2002.pdf.

5. Barbara Bardes and Robert W. Oldendick, *Public Opinion: Measuring the American Mind*, 4th ed. (Lanham, MD: Rowman & Littlefield Publishers, 2012).

6. "Records of the Bureau of Agricultural Economics [BAE]," National Archives, last modified August 15, 2016, https://www.archives.gov/research/guide-fed-records/groups/083.html.

7. Andrew Mercer, Claudia Deane, and Kyley McGeeney, "Why 2016 Election Polls Missed Their Mark," Pew Research Center, November 9, 2016, http://pewrsr.ch/2fmNGH6.

8. Mona Chalabi, "Yes, the Election Polls Were Wrong. Here's Why," *Guardian*, November 9, 2016, https://www.theguardian.com/commentisfree/2016/nov/09/polls-wrong-donald-trump-election.

9. William Lunch and Peter Sperlich, "American Public Opinion and the War in Vietnam," *Western Political Quarterly* 32, no. 1 (1979): 21–44, http://doi.org/10.2307/447561.

10. "Confidence in Institutions," Gallup, accessed July 23, 2020, https://news.gallup.com/poll/1597/confidence-institutions.aspx.

11. John B. Horrigan, "The Elements of the Information-Engagement Typology," Pew Research Center, September 11, 2017, https://www.pewresearch.org/internet/2017/09/11/the-elements-of-the-information-engagement-typology.

12. R. J. Reinhart, "Nurses Continue to Rate Highest in Honesty, Ethics," Gallup, January 6, 2020, https://news.gallup.com/poll/274673/nurses-continue-rate-highest-honesty-ethics.aspx.

13. "Public Trust in Journalists Is Down, But in 'Media' It's Up," News Literacy Project, accessed July 23, 2020, https://newslit.org/get-smart/did-you-know-trust-in-journalists-media.

14. Ibid. Edelman defines the trust index, the numbers I am relaying, as the average derived from what percent of people trust nongovernmental organizations, the media, government, and business.

15. Roderick M. Kramer, "Rethinking Trust," *Harvard Business Review*, June 2009, https://hbr.org/2009/06/rethinking-trust.

16. Ibid.

17. Meghna Chakrabarti and David Folkenflik, "Are We Living in a Post-Truth World?" *On Point*, produced by WBUR for NPR, February 27, 2020, https://www.wbur.org/onpoint/2020/02/27/part-iv-post-truth.

18. Nick Routley, "Who Owns Your Favorite News Media Outlet?" *Equities News*, October 7, 2019, https://www.equities.com/news/who-owns-your-favorite-news-media-outlet.

19. Jeffrey M. Jones, "U.S. Church Membership Down Sharply in Past Two Decades," Gallup, April 18, 2019, https://news.gallup.com/poll/248837/church-membership-down-sharply-past-two-decades.aspx.

20. "Public Library Use: ALA Library Fact Sheet 6," American Library Association, last modified October 2015, http://www.ala.org/tools/libfactsheets/alalibraryfactsheet06; Lee Rainie, "The Information Needs of Citizens: Where Libraries Fit In," Pew Research Center, April 9, 2018, https://www.pewresearch.org/internet/2018/04/09/the-information-needs-of-citizens-where-libraries-fit-in.

21. "Libraries: How They Stack Up," OCLC Online Computer Library Center, Inc., accessed November 28, 2015, https://www.oclc.org/content/dam/oclc/reports/librariesstackup.pdf.

22. Data on public library usage comes from the annual Institute of Museum and Library Services report, which includes data on the over 18,000 public libraries in the United States. The latest data is from 2017. "Public Libraries Survey," Institute of Museum and Library Services, accessed July 23, 2020, https://www.imls.gov/research-evaluation/data-collection/public-libraries-survey.

23. "Publications." ALMPUB, accessed July 23, 2020, https://almpub.wordpress.com/publications.

24. Eric Klinenberg, *Palaces for the People: How Social Infrastructure Can Help Fight Inequality, Polarization, and the Decline of Civic Life* (New York: Broadway Books, 2019).

25. "Times Topics: Libraries and Librarians," *New York Times*, accessed July 23, 2020, https://www.nytimes.com/topic/subject/libraries-and-librarians; Eric Klinenberg, "Palaces for the People: Why Libraries Are More Than Just Books," *Guardian*, September 24, 2018, https://www.theguardian.com/cities/2018/sep/24/palaces-for-the-people-at-the-library-everyone-is-welcome; Mick Krever, "Imagination Under Threat," Amanpour, October 30, 2013, http://amanpour.blogs

.cnn.com/2013/10/30/imagination-under-threat; Kalev Leetaru, "Computer Science Could Learn a Lot from Library and Information Science," *Forbes*, August 5, 2019, https://www.forbes.com/sites/kalevleetaru/2019/08/05/computer-science-could-learn-a-lot-from-library-and-information-science/#15b0e06a587d.

26. "Traditional Families Account for Only 7 Percent of U.S. Households," Population Reference Bureau, March 2, 2003, https://www.prb.org/traditional-families-account-for-only-7-percent-of-u-s-households.

27. R. David Lankes, *The New Librarianship Field Guide* (Cambridge, MA: MIT Press, 2016), 3.

28. Frank Newman, Lara Couturier, and Jamie Scurry, *The Future of Higher Education: Rhetoric, Reality, and the Risks of the Market* (New York: John Wiley & Sons, 2010).

29. Rachel Fishman, Sophie Nguyen, Alejandra Acosta, and Ashley Clark, "Varying Degrees 2019," New America, last modified September 10, 2019, https://www.newamerica.org/education-policy/reports/varying-degrees-2019.

30. Scott Jaschik, "Falling Confidence in Higher Ed," *Inside Higher Ed*, October 9, 2018, https://www.insidehighered.com/news/2018/10/09/gallup-survey-finds-falling-confidence-higher-education.

31. Ibid.

Media Waypoint

The Weaponized Knowledge Infrastructure

When I was a child, I remember walking through a natural history museum exhibit on the Ice Age. There were large statues of wooly mammoths and saber-toothed tigers. Displays on the wall marked the progression and retreat of glaciers across the Earth's surface. Models showed the formation of the Great Lakes.

I distinctly remember one poster that asked, "What was it like to live in an ice age?" The answer below was somewhat startling—you are living in one now. The northern ice cap and the frozen Antarctic are holdovers from the last advance of the ice (well, at least for a few more years).

The point was clear: oftentimes we can't see the reality of an era or epoch around us. Either it is so enmeshed in our current environment, or it is discussed as part of the past, or an issue elsewhere. I don't need to ask what it is like to live in a knowledge infrastructure that has been weaponized and is constantly being shaped by war and conflict—you are living in one now. Here are just a few examples of how you are living in a time when information is used to increase anxiety around conflict.

On March 11, 2002, the newly created Department of Homeland Security unveiled the Homeland Security Advisory System. The system had a series of colors to represent the likelihood of a terrorist attack: from green (low) to blue (guarded) to yellow (elevated), orange (high), and red (severe). From its creation until it was replaced on April 27, 2011, with a new system called the National Terrorism Advisory System, the alert level never fell below yellow.

In 1976 Congress passed the National Emergencies Act to formalize (and constrain) the emergency powers of the president. As of this writing, nine presidents have declared 60 emergencies since the inception of the law. Thirty-three of these are still in effect. They range from sanctions on Iran for "Proliferation of Weapons of Mass Destruction" to "Blocking Assets and Prohibiting Transactions with Significant Narcotics Traffickers" to "Blocking Property of the Government of the Russian Federation Relating to the Disposition of Highly Enriched Uranium Extracted from Nuclear Weapons."

In the United States, since 1948 men have been registering for a possible draft once they turn 18. Telephone carriers have been sharing metadata on domestic calls to counter terrorism since the adoption of the Patriot Act in 2001, and the National Security Administration regularly monitors calls made to international numbers. The war in Afghanistan, the longest war in U.S. history, has been fought since 2001. We fight wars on terror, on poverty, AIDS, cancer, and drugs.

Media manipulation is not just an historical artifact of World War I; it is happening today. In 2016 the U.S. government spent more than a billion dollars on public relations and advertising. A significant portion of that is spent by the military: "The Pentagon accounted for 60 percent of all public relations spending between 2006 and 2015, the GAO found, and it employs about 40 percent of the more than 5,000 public relations workers in the federal government."[1] This includes paying for patriotic (and military) displays at sporting events.[2] Is this money going to calm the public and make people feel secure? Do you feel safer? Are you safer? It turns out that how safe you feel and how safe you are have two different answers.

Once again, looking at the data—the data analyzed, qualified, and reviewed, as opposed to the data that is at hand to crunch—we see ample evidence that globally, we are living in the safest decade in recorded history. Max Roser, an economist at Oxford University, created a site called ourworldindata.org that demonstrates our relative calm and peace with a number of factors.

Take deaths in war. Since 1947 the high point for deaths was 1950, with 546,501. Deaths also spiked in 1972 with the Vietnam War at 291,861. As late as 1984, war deaths were at 237,032. In 2016, the last date presented? 87,432. Fewer people are dying in war. Take a look in figure 10.9.

Battle-related deaths in state-based conflicts since 1946, by world region , 1946 to 2016

The region refers not to the location of the battle but to the location of the primary state or states involved in the conflict (see 'Sources' tab). Only conflicts in which at least one party was the government of a state and which generated more than 25 battle-related deaths are included. The data refer to direct violent deaths (i.e. excluding outbreaks of disease or famine).

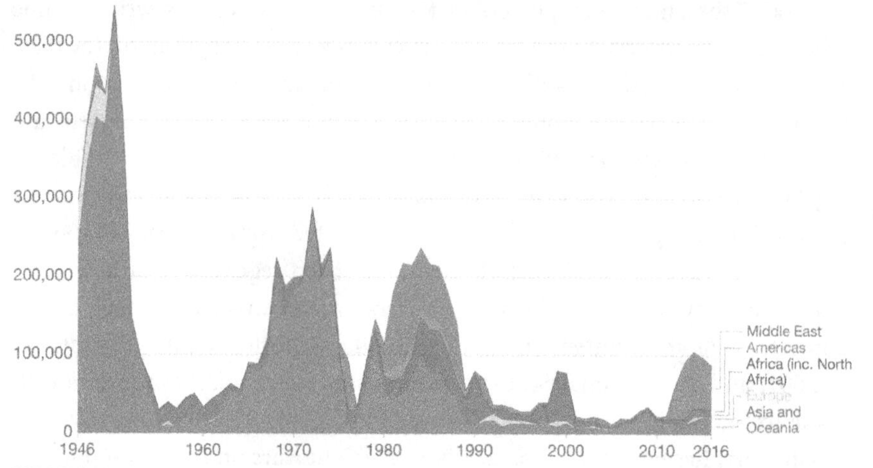

FIGURE 10.9

Battle-related deaths in state-based conflicts since 1946, by world region, 1946 to 2016. ***Max Roser, "War and Peace," Our World in Data, 2016, https://ourworldindata .org/war-and-peace***

As Christopher Fettweis, author of *Dangerous Times? The International Politics of Great Power Peace*, puts it:

> Although it may seem counterintuitive to those whose historical perspective has been warped by the twenty-four-hour-news cycle, levels of conflict, both in terms of number and magnitude, have been dropping steadily since the end of the Cold War. A series of empirical analyses done in the United States and Canada have consistently shown that the number of wars of all types—interstate, civil, ethnic, revolutionary, etc.—declined throughout the 1990s and into the new century. The risk for the average person of dying violently at the hands of enemies has never been lower.[3]

There has been a steady decline in global poverty as well: from 94 percent of the world living in poverty in 1820 to 52 percent in 1992.[4] Homicide rates across the globe have been declining as well; the really long-term picture since the 1300s in European countries is shown in figure 10.10.

But a little closer to home, in the United States, the trend is real. Since 1990 both violent crime and property crime have been falling to the point where

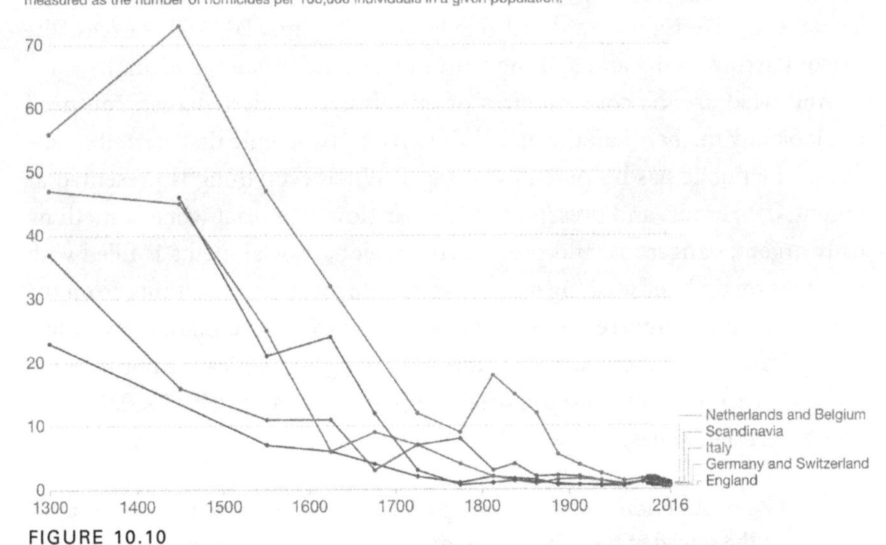

FIGURE 10.10
The stark decline in homicides in the west since the 1300s. *Max Roser and Hannah Ritchie, "Homicides," Our World in Data, July 2013, https://ourworldindata.org/war-and-peace*

there is roughly half the crime today as a quarter-century ago.[5] At the same time, more people *believe* that the crime rate is rising.

A hungry 24-hour news cycle and political campaign rhetoric continuously seek stories of conflict and war. Political coverage is dominated by the "horse race," or poll numbers and who is winning, over in-depth coverage of policy positions and proposals. A weaponized knowledge infrastructure driven by the need for greater ad revenue or votes seeks out winners and losers over common ground and movement forward. Nationalism has replaced globalization as a leading national narrative and, with it, rising cases of xenophobia and racism. The resurgence of racism can be seen in everything from an increase in campus hate crimes,[6] to people's perception of discrimination,[7] to police killings of Black Americans.

Throughout this "Media" section we have seen increased fragmentation and how our institutions, like the media and government, may actually drive this. We can see how this division is encouraged by all aspects of the knowledge infrastructure from people (separation into ideological tribes driven by confirmation bias), technology (data analysis and AI development without recognizing implicit and explicit bias in our tools), policies (expansive intellectual property rights, outdated regulation of the media), and sources (the rise of narrowcasting and shifting trust in institutions like the media).

And what are the consequences of this always-on-alert status? You need not look any further than the global COVID-19 pandemic that started in late 2019. The public has become desensitized. When everything is presented as urgent, dangerous, and present, they are far slower to react when something truly urgent, dangerous, and present comes along. Social media is filled with information on handwashing and conspiracy theories on everything from the virus being an engineered weapon to an excuse to declare martial law. Messaging from government scientists filtered through a political response led to a lack of trust in urgent and important messaging, as a March 17, 2020 NPR/PBS Newshour/Marist poll found:

> Just 46% of Americans now say the federal government is doing enough to prevent the spread of the coronavirus, down from 61% in February.
>
> Just 37% of Americans now say they had a good amount or a great deal of trust in what they're hearing from the president, while 60% say they had not very much or no trust at all in what he's saying.[8]

Citizens confined to homes and within closed national borders turn to a media unable to find the fine line between raising awareness and inducing panic.

Pandemic measures of social distancing are pushing people increasingly to a knowledge infrastructure tuned to make money over a need to make sense. News channels are interspersing handwashing demonstrations with pundits calculating which political party will benefit from a higher or lower death toll. Patriotic swells seek to honor our front-line health workers and still extol "American Exceptionalism" even as the exception that the American response demonstrates is rising infection and mortality rates. Vaccine development is already hampered by a system that allows the antivaxxer narrative equal time with life-saving science.

We need a serious rethinking of the knowledge infrastructure. We need a new model of developing our worldview from a place of learning, humanity, and rationality. In the next section I will seek to build just such an agenda. In order to do so I will use the early days of the COVID-19 pandemic as a case study and launching point for arguing reform. I argue that this global phenomenon demonstrates what needs to change, and also highlights why a rush to dataism isn't a solution.

NOTES

1. Eric Boehm, "The Pentagon Accounts for More Than Half of the Federal Government's $1 Billion PR Budget," *Reason*, October 10, 2016, https://reason.com/2016/10/10/the-pentagon-accounts-for-more-than-half.

2. Eyder Peralta, "Pentagon Paid Sports Teams Millions for 'Paid Patriotism' Events," *NPR*, November 5, 2015, https://www.npr.org/sections/thetwo-way/2015/11/05/454834662/pentagon-paid-sports-teams-millions-for-paid-patriotism-events.

3. Christopher J. Fettweis, "The World Is Actually Safer Than It Used to Be . . . And It Keeps Getting Safer," History News Network, accessed July 29, 2020, https://historynewsnetwork.org/article/142159.

4. Max Roser and Esteban Ortiz-Ospina, "Global Extreme Poverty," Our World in Data, last modified March 27, 2017, https://ourworldindata.org/extreme-poverty.

5. Dara Lind, "The US is Safer Than Ever—and Americans Don't Have Any Idea," Vox, last modified April 7, 2016, https://www.vox.com/2015/5/4/8546497/crime-rate-america.

6. Victoria Nelson, "Addressing Racial Trauma and Hate Crimes on College Campuses," Center for American Progress, August 9, 2019, https://www.americanprogress.org/issues/race/news/2019/08/09/473299/addressing-racial-trauma-hate-crimes-college-campuses.

7. Samantha Neal, "Views of Racism as a Major Problem Increase Sharply, Especially Among Democrats," Pew Research Center, August 29, 2017, https://www.pewresearch.org/fact-tank/2017/08/29/views-of-racism-as-a-major-problem-increase-sharply-especially-among-democrats.

8. Domenico Montanaro, "Poll: Americans Don't Trust What They're Hearing from Trump on Coronavirus," NPR, March 17, 2020, https://www.npr.org/2020/03/17/816680033/poll-americans-dont-trust-what-they-re-hearing-from-trump-on-coronavirus.

Part III

SOCIETY

Activism is the rent I pay for living on this planet.

—Alice Walker

We don't accomplish anything in this world alone . . . and whatever happens is the result of the whole tapestry of one's life and all the weavings of individual threads from one to another that creates something.

—Sandra Day O'Connor

11

Certainty

From Clockwork to Complexity

The knowledge infrastructure is the way we come to know the world to find meaning and power in it. The infrastructure lies at the intersection of data and media—what we know, and the channels we use to know it. This infrastructure has been shaped over the past century by war and conflict. Military investment and nations' ambitions to attain power and territory have led to the development of encryption, ubiquitous data collection, massive-scale computing, and propaganda. It has also led to a knowledge infrastructure that prizes certainty and uses the prospect of uncertainty to create anxiety (will there be another terrorist attack? Will the coronavirus kill us all?).

I argue that development and reform of the infrastructure must be based on a humanist/rational approach as opposed to the growing dataism or, though not explicitly discussed, other belief systems like religion or nationalism. In essence. I'm not saying we need to change the knowledge infrastructure because God wants us to, or so that the United States can be great again. Rather, I'm saying we need to change it so we can better find meaning and personal power in today's society. What I am arguing for is a new frame of reference.

In seeking a better path through improving the knowledge infrastructure—one unencumbered by a violent past—it is important that we understand the effects of this knowledge infrastructure on our lives today. Without such context, any proposal for change will seem arbitrary, or even unnecessary. What is needed is a sort of case study—some revelatory moment that we can all identify with, one that highlights the importance of a functional knowledge infrastructure in society. One that affects everyone across all sectors of society: across the political spectrum, across class, across race, across educational attainment. Some event where we all must struggle to find meaning and power through knowledge because neither physical force nor financial well-being will affect the outcome. Something like a global pandemic.

This book was written in the middle of 2020, when the pandemic was still months old. While later books and no doubt innumerable articles and media spots will seek to write the definitive history of the pandemic, it is the initial days that are most relevant to this work. It is in the initial days where information is most crucial, action is most needed, and the stress on the knowledge infrastructure is greatest.

I will use the early and often chaotic first months of the pandemic—roughly from the beginning of the first wave of the virus, through the lockdown, and the beginning of the initial reopening of the economy—to demonstrate the dangers of a weaponized knowledge infrastructure. Over the next chapters, I turn my historic focus from a century to 31 days in March 2020. I will use this "history of the present" to not only demonstrate shortcomings of the infrastructure, but to argue for ways it can be improved. If the global COVID-19 pandemic is truly a world-changing event, then we had better take full advantage of this horrific opportunity to ensure change for the better.

In the opening months of the pandemic's spread throughout the United States, a huge percentage of the global population stayed home. Businesses, schools, museums, government offices, and library buildings were closed. Restaurants that were not closed could only offer take-out or delivery. Store shelves were bare of essentials like toilet paper and milk, and supermarkets were rationing meat and other products. Government and commercial employees alike were told to work from home or were laid off. In the United States, over 300 million people were in lockdown.

National borders were closed across the globe, including the European Union, which busily rebuilt checkpoints closed by the 1985 Schengen Agreement. In Jordan, Italy, and Spain, citizens were forced to isolate in their homes. Italian doctors faced with severely limited emergency medical equipment had to pick who lived and who died. It was not a time of certainty, to say the very least.

The State of New York was the epicenter of the pandemic on March 3, 2020, with 5 percent of the global virus cases. Governor Andrew Cuomo had shuttered schools, bars, and any gathering over 10 people—the lights of Broadway were dark. He directed the National Guard to confiscate ventilators and unused medical supplies from upstate counties with few cases for use in hard-hit New York City. The U.S. Army erected field hospitals in parking lots and Central Park. U.S. Navy hospital ships *Mercy* and *Comfort* were deployed to Los Angeles and Manhattan respectively as the death count rose.

The stock market had plunged, losing trillions in value. Congress had just passed a rescue economic package with a price tag of over 9 percent of the country's Gross Domestic Product.

As the novel coronavirus spread through the United States in March, President Trump offered shifting and sometimes dangerous narratives: "It is just like the seasonal flu," or it "would go away in the heat," or "The country needs to open for business by Easter." In the early days of the pandemic, data acted as a sort of antidote to presidential confusion. The two recognized bright spots of the national response to the pandemic were Dr. Anthony Fauci and Dr. Deborah Birx. Both had deep backgrounds in epidemiology and infectious disease. Both were persistent reporters and interpreters of data.

Dr. Birx referred to the collected data of infection and death rates as her sheet music—using it to guide advice. Dr. Fauci regularly quashed treatment discussions and forecasting with the need for real data and clinical trials. Both noted the importance of testing and the need for better contextualizing data.

At that time, estimates of U.S. deaths from the virus ranged from 200,000 to 1.7 million.[1] These estimates came from data-intensive scientific models. Each model, however, varied in its assumptions (number of reported infections vs. the actual number, how compliant a population will be toward social distancing, etc.) and in which variables were considered and how these variables are weighted in importance. In March, according to the Centers for Disease Control and Prevention (CDC) and Dr. Birx, the only data that had some reality were the mortality counts. The model that projected 100,000–200,00 deaths on the "optimistic" side? It made a series of assumptions such as: all states will issue stay-at-home orders (by the end of March only 33 had); and that all orders will stay in place until June 1 (the president issued social distancing guidelines until the end of April).[2] If these guidelines were not followed, the death rate was projected to be much higher. How many people would die from this pandemic? As Matt Finch, an expert in scenario planning, once told me, "you can't gather data from the future."[3]

So, while data may well be a great antidote to wild speculation, it is not the bastion of certainty it appears. Which brings us to where I try and convince you that uncertainty is just as important as certainty.

In this chapter I will show how science has evolved over the past century to provide us with guidance to advance society's knowledge infrastructure. I will also show how the evolution of science demonstrates why a search for certainty has fundamentally shifted away from universal solutions, and in doing so, invalidates the underlying certainty and determinism that dataism appears to offer. To do that—you guessed it—I turn to history and to the *Académie des Sciences* in Paris in 1798.

FROM LAPLACE'S DEMON TO EINSTEIN'S CLOCKS

To say Pierre-Simon Laplace had an effect on the world of mathematics is a little like saying Thomas Edison had an effect on the lighting industry. Laplace virtually introduced mathematics into the field of physics and laid the foundation for the study of heat, magnetism, and electricity. He built on Newton's work to mathematically explain the motion of objects in the solar system. This was no small feat. as even Newton said the problem was so complex that "divine intervention was periodically required"[4] to make it work.

Laplace is credited for developing a philosophy of science known as *determinism*. He summed it up best when he wrote:

> We may regard the present state of the universe as the effect of its past and the cause of its future. An intellect that at any given moment knew all the forces that animate nature and the mutual positions of the beings that compose it, if this intellect were vast enough to submit its data to analysis, could condense into a single formula the movement of the greatest bodies of the universe and that of the lightest atoms; for such an intellect nothing could be uncertain; and the future, just like the past, would be present before its eyes.[5]

These are Laplace's words, but they could easily be the words of a data scientist in Silicon Valley, or an analyst at the Pentagon today. They are an elegant encapsulation of both the ethos of the Enlightenment and the rising dataism of today. They represent a fundamental distrust of uncertainty, and a belief in a deterministic and mechanical universe. Not to put too fine a point on it, but it is also a pretty concise definition of God. Except here God is not a deity or a supernatural being, but the ultimate evolution of human intellect—an intellect founded in logic and empiricism.*

Looking back over the past century, one could easily see it as a march toward certainty—a quest to build into the knowledge infrastructure precision and predictive power referenced by Laplace. The military developed the internet and computers to eliminate uncertainty on the battlefield. Advertisers developed psychological means to precisely target the right consumer to assure sales. Politicians looked to polling to craft just the right message for the next election. Today we embed chips in asphalt to control the environment, and we store faces in databases to know what people are doing from one setting to another. Our watches no longer just tell the time, but rather use microscopic gyroscopes to identify our unique pattern of typing and monitor the beats of our hearts. Shadow digital personas mimic our desires, not for self-exploration, but for monetization.

Nowhere is this drive for certainty and predictive power on better display than in the sciences, from the physical sciences' quest for the theory of everything in the cosmological and sub-atomic realm to the efforts of social sciences like psychology, sociology, and education to see through the seeming randomness of social interaction and human desire. Yet, a funny thing happened on the way to a world of flying cars and social rationalism: the universe played dice.

When the CS *Alert* set off early on that August morning in 1914, it sailed into a universe in flux. To be precise, the human view of the universe was in flux—the universe probably didn't care. Up until the end of the nineteenth century, most of how we understood the physical world, the field of physics, was dominated by classical mechanics. This model of the universe (what I'll call a paradigm in a moment) traces its roots all the way back to Isaac Newton and his now famous (and most likely apocryphal) apple. His equations defined an orderly world, and his equations of gravity still play a role in everything from model rockets to actual rockets shooting into space.

This mechanical view of the world dominated physics throughout the nineteenth century. That's not to say Newton had figured everything out. Over the next two centuries, theories to explain everything from optics to magnetism to electricity (well, at least static electricity—it's complicated)

* The idea for this vast intellect is known as Laplace's demon.

to chemistry were added to the understanding of how the physical world seemed to operate. This was an amazing time in science as it seemed new discoveries about the fundamentals of the universe were being revealed and explained at a breathtaking pace.

The eighteenth and nineteenth centuries in Europe and North America generated such a new view of how the world operated that it became known as the Enlightenment. Not since the Renaissance had the western world experienced such a rapid and radical shift in the world of thought. The appearance that humankind was now unlocking the clockwork of nature extended well beyond physics. In fact, the new ability of the human mind to understand and explain nature led to the codification of humanism. As scientific explanations replaced mystical and religious dogma, humans were seen increasingly at the center of the universe . . . even though science made it clear, the Earth was not.

To be honest about history, however, it would be more precise to say western, white, elite, and primarily male humans were putting themselves at the center of the universe. The new keys to unlocking the universe engendered a belief in the unique role European culture held in the concept of civilization. It meant that the "backward" views of other peoples and nations signaled less refinement and development. The mastery of the physical world through science seemed ample evidence that Europe should master the rest of the world through imperialism, racism, and violence. "Ironic" seems way too small a word to describe how a philosophy about the ascendance and primacy of the individual was used to enslave other individuals. "Barbaric" is better.

To be clear, the subversion of other cultures through technological and military domination was not invented in the Enlightenment, nor was it the first time that philosophy was put in service of such domination. Yet the fact that humanism, a concept I am arguing for, was invented and used to justify the suppression of the agency and power of entire nations is an important reminder as we talk about retaking the knowledge infrastructure.

By the start of the twentieth century, however, the Enlightenment was coming to a close both in terms of science and imperialism. As more and more colonies broke with their imperial masters, the mechanical universe of Isaac Newton and James Maxwell was beginning to break down as well.

A core tenet of the scientific process and underlying worldview is *falsifiability*.[6] A scientific theory (like the theory of evolution or gravity) must be able to be tested and proved wrong. This is different from religion, where the basic belief that God exists cannot be definitely proven wrong. "I believe there is a God" is a statement that doesn't lend itself to experimentation. "If I drop this bowling ball it will fall at 32 feet per second per second" (not a typo) does. This is why I say that science is a pursuit of the truth that never claims to have discovered it. All of science is conditional. It is the truth until any evidence that doesn't fit with the explanation shows it is not. By the time the *Alert* sailed, the evidence was mounting that the mechanical view of the universe, and the certainty it provided, was false.

The mechanical view of physics couldn't explain the newly discovered radiation pioneered in the work of Marie and Pierre Curie. It also couldn't explain how the immutable atom could actually itself consisted of things like electrons, protons, and neutrons. Worse still, while concepts like Newton's gravity and concepts like a universal ether that allowed light to move through seemingly empty space worked at the large scale, it was failing again and again in new experiments at the microscopic and atomic scales.

Suddenly, the certainty of the Enlightenment was in peril. Thank God for Albert Einstein.

FROM RIDING A LIGHT BEAM TO RIDING A PHOTON

It was the search for a new certainty that led Albert Einstein to propose his special and general theories of relativity and change the world. His theories elegantly explained all of the measurements that didn't fit in the mechanical theories. He replaced strained concepts such as a "luminiferous ether" with concepts such as distortion of space-time and the speed of light as a constant, and introduced new ways of understanding the universe without a single fixed perspective. While the speed of light in a vacuum is constant, time and space itself can change from the perspective of the viewer.

In this model—the dominant model of physics at the outset of World War I—it wasn't the viewer that changed time or space, but simply the frame of reference. This differed from quantum mechanics, the theory that sought to explain how the universe operated at the subatomic scale. Still, under the theory of general relativity, physics could return to the business of certainty. Equations could be written and solved to determine the motion of planets and bowling balls dropped from tall towers. While it might be hard for many of us to wrap our head around things moving near the speed of light, we have all seen the implications of one of Einstein's equations: $E=MC^2$. In bringing certainty back to physics, Einstein (and, to be fair, thousands of other physicists) tied light and energy and matter together with gravity and showed us a universe where what we thought of as invariant, like time and space, were dynamic. Dynamic to the point that 141 pounds of uranium could level Hiroshima.

When general relativity also began to fail to explain new physical phenomena, a new model of the universe was developed: quantum mechanics. Just as relativity represented a new fundamental way of thinking about the universe (from a fixed and static space to one dependent on frames of reference and a fabric of space and time that distorted in the presence of matter), quantum physics radically altered this view again—from a macroscopic world of unified forces to a set of discrete units of energy and matter that acted both as particles and waves.

Whereas in relativity an observer's frame of reference changes perception, in quantum mechanics the act of observation itself could literally change the state of a particle. Rather than a clockwork, deterministic universe, quantum mechanics was all about probabilities and randomness. The stability and predictability of the world around us was now the result of a near-infinite number of chance events coalescing into the everyday. Probabilities replaced coordinates, and electron clouds replaced the sleek ringed orbits we were shown in high school.

But here's the thing: Newtonian mechanics, Einsteinian relativity, and quantum theory all still play a role in how science and engineering work today. We fly planes and crash test dummies in cars with classical physics—the equations still work at "normal" speeds and sizes. The GPS that gives you directions in your car is still proving Einstein right. And we wouldn't be able to make those ubiquitous chips with billions of transistors without an understanding of quantum mechanics. These systems all seem to work, even though they are not fully compatible.

Still, many continued to buy into the myth that science was a continuous path getting ever-closer to the truth and certainty. This was the myth that Thomas Kuhn blew up in his 1962 seminal work, *The Structure of Scientific Revolutions*.[7]

Kuhn was a philosopher of science who studied the way bodies of thought would evolve, and ultimately be superseded by, new schools of thought. In his research, he noted that while the predominant narrative was that science (he focused on physical sciences) proceeded with one discovery after another, smoothly adding bricks of discovery onto the wall of understanding, the truth was far more complex. He used the word "paradigm" to talk about how schools of scientific thought coalesced. He argued that people in a paradigm shared common beliefs, approaches, and blind spots. The prevailing "mainstream" paradigm reified itself with scholars indoctrinating their students, who became professors that indoctrinated their students, and so on. Think of it as the original filter bubble.

He found that these paradigms and their often-unstated beliefs were virtually invisible to those in the paradigm. They didn't see cracks in their theories or missing data because they never even thought to ask questions that would show them. It would take people with a new perspective who operated on the fringes of the paradigm to see the gaps and then propose solutions. These outsiders might be new to the profession, or come from other fields altogether.

Once the outsiders demonstrated a new way—Newton with gravity; Einstein with relativity; and Max Born, Werner Heisenberg, and Wolfgang Pauli with quantum mechanics—they move to the center, displacing the old paradigm with the new. The evolution of science wasn't a linear march to certainty and truth; it was more an episodic battle of ideas, all professing a better worldview.

In his later years, Kuhn tempered his views on paradigms. He and others refined the idea, bringing in more nuanced views of scientific revolutions and evolutions. Yet the idea remained: the universe refused to yield to a universal clockwork in which existed, as Laplace in 1798 had longed for, an intellect for which nothing could be uncertain.

FROM QUANTUM PHYSICS TO THE WINGS OF A BUTTERFLY

As in the days of the Enlightenment, the new complex view of the physical universe influenced other fields of science and scholarship. Fields such as psychology, sociology, anthropology, literature, philosophy, and mathematics went through similar moments of evolution and, in some cases, crisis as new concepts of certainty and determinism were challenged.

Social science at the beginning of the twentieth century was well on its way, or so it thought, to unlocking universal laws for human behavior. Gallup and the development of scientific polling is an example I've already described. In World War II, operations research developed as a field, first seeking to improve the accuracy of radar and bombing, and then turned into all things "efficient." To say some operations research scholars had grand ambitions might be underselling it a bit:

> In 1945 the British crystallographer and Marxist intellectual J. D. Bernal went so far as to suppose that wartime O.R. [operations research] represented not a new profession, but a total realignment in the relations between science, the state and society. He reckoned that the moment marked the beginning of an entirely new epoch in history in which human progress could be intelligently planned.[8]

Fields like operations research and the growing field of management science and economics were in search for universal truths. Then Herb Simons came along and won a Nobel Prize in economics for his concept of "satisficing."

Simons was wrestling with how people within an organization make decisions. The general idea (and to be clear, I am not doing it justice) was that all types of organizations want to optimize their systems—to make the best decision, whether those systems are for shipping goods, managing human resources, or controlling world economies. Many economic theories and those from management science were premised on perfect knowledge, that is, making the best decision with the best information. Simon pointed out that no organization or individual can ever have perfect knowledge. A company never knows, for example, what a competitor will do. An investor never knows what a market will do if, say, a novel coronavirus spreads into a pandemic. So, people make decisions that they anticipate will bring the best result for the least amount of resources risked.

Satisficing as a concept has been widely adopted in everything from finance to education. For our purposes, satisficing is an effective marker when many fields rejected the idea that everything could be predicted. This core concession would go on to whole new ways of thinking about systems in general.

In fact, perhaps the biggest transformation was toward something known as general systems theory. The theory was a way to conceptualize systems from how the human brain worked to how an ecosystem could be maintained. You have probably run into it, even if not by name. General systems theory says that a system consists of inputs, processes, and outputs. It forms the basis of processes such as assessing the math knowledge of high school students. What is the outcome you want to achieve, what process do you need to put in place for that outcome, and what data can you use to assess whether the outcome was achieved? The problem was that a system like this was akin to perfect knowledge. In practice, what inputs were available and how an effect is measured is, well, complicated. In fact, it's complex.

The first attempts to save general systems theory was to realize that the output of a system needed to also become part of the input: a feedback loop. How could you know if the process you put in place to teach math in high school was working if you didn't test and then refine the process? This addition was called cybernetics. Then—and you had to see this coming—cybernetics ran into problems. Both cybernetics and general systems before it assumed a closed system. Take an ecosystem like a forest. In both general systems and cybernetics, that ecosystem was assumed to be self-contained. The amount of water available, the species of animals using it as a home, the flora that grew, and other factors were fixed—like a terrarium in a jar. But what about invasive species? What about the fact that the water system you are looking at is tied to the country next door? Climate change, for example, is forcing certain species to go farther afield for food and certain plants to bloom earlier, crossing what were once thought of as fixed boundaries.

It gets more complicated still. Up until the 1970s, most of systems theory worked in linear models. All of those processes assumed that the impact of all the inputs could be modeled and accounted for. With the introduction of chaos theory and complexity theory, that assumption was disproven—small changes in a system could have massive and unanticipated effects. A complicated problem or system is one with a lot of parts. Yet each part could be identified, and the impact of that part to the system as a whole could be predicted. Complicated problems took a lot of work to understand, but they could be understood. Complex problems or systems, on the other hand, were fundamentally different. Try as you might to identify every part and its contribution, that contribution not only might have larger effects than you anticipate, but the component and the impact were dynamic—they could change over time as they interacted with other parts.

The easiest way to explain why is known as the "butterfly effect." The normal explanation of the butterfly effect is in weather. A butterfly flapping its wings does not generate a lot of wind. Yet it can add a minuscule amount to the velocity of wind in, say, Brazil. That microscopically accelerated wind can push an approaching high-pressure ridge microns forward, just enough to merge with another front moving across Brazil, that can merge with a high-altitude jet stream across the Atlantic Ocean, that can create a vortex of air over rising ocean temperatures, that can form a hurricane.

However, a better example might be to talk about all of those contagion models that epidemiologists are using to predict deaths and infections in the global pandemic discussed earlier. These models try to account for as many variables and use as much processing power as they can to create better predictions. To improve the models, new variables are added all the time. Over time, these models become more precise and more complicated . . . they account for more inputs. However, they struggle to become more complex.

The first problem is that no one will ever find all the variables. Does humidity or temperature matter in transmission? The second problem is that those variables can never capture the randomness of reality. RNA viruses like the novel coronavirus are constantly mutating—will one of these mutations lead to a more virulent or more indolent strain? The third problem is that models don't do well with small variables having outsized effects. My son gets the virus, he is young and may have few health issues. He is also spending a lot of this pandemic in his room playing with video games and meeting with friends and classmates online. That's very different than, say, if the son of my college dean gets the infection. The son may infect his father, who is considered an essential employee and so must go to campus, where he may meet with the university president, who meets with the state's governor, who may meet with the president of the United States. In these models there is no level of precision that will predicate different outcomes based on two males under 20 in the midlands of South Carolina.

Before I leave this example, let me say that these models are good. They are useful and they will help save lives. However, they are also currently, and always will be, works in progress. As with weather modeling that has greatly improved hurricane prediction, more computing, more variables, and the introduction of machine learning and cleaner data will make these models better. But let us not for a moment forget that the science, experience of the modelers, and underlying epidemiological theory are fundamental to improved outcomes. Or, put more succinctly by Dr. Anthony Fauci: "Whenever the models come in, they give a worst-case scenario and a best-case scenario. Generally, the reality is somewhere in the middle. I've never seen a model of the diseases that I've dealt with where the worst case actually came out. They always overshoot."[9]

Once mathematicians began looking around at the world, they found all sorts of examples of small changes making massive changes. Everything from the stock market to storm patterns on Jupiter to whether or not a video goes viral on social media seemed to behave in a similar fashion. Chaos theory,[10] and later complexity theory,[11] gave up the idea of determining the behaviors of such systems and moved toward modeling them instead. The result of their models and calculations were less about predictions of certainty than about seeking patterns and possibilities. Systems were no longer inputs, processes, and outputs, but a world of autonomous agents detecting things in the environment, building heuristic rules to process that information, and then effectors to try and change the environment in their favor. However, the link between detectors and effectors was not straight. Rules could evolve and change.

Just as in physics, where atoms were no longer single entities but a complex and, to an extent, unpredictable combinations of fundamental particles, systems were agents interacting in complex, and often unpredictable, ways. In sociology and economics, models of people as rational agents making optimal decisions were replaced by complex models and predictions bounded by uncertainty. Who could have predicted at the outset of the pandemic, for example, that some people would refuse to wear masks, and that this refusal was associated with political ideology?

It is this paradigm, in which small efforts can have outsized effects, that stands at the core of humanism. Your actions can influence 2 other people, who can change the behavior of 4 people, then 16, 32, 64, 128, and on and on exponentially. One sick person can infect two who infect four, and on and on until we have a global pandemic. In a complicated world, where a lot of small parts have a deterministic and linear effect, the power of the individual will always be muted and the examination of such a world is centered on data from a lot of individual sources seeking a cumulative effect. In a complex world? Well, as the saying goes, "Never doubt that a small group of thoughtful, committed citizens can change the world. Indeed, it is the only thing that ever has."*

This cycle of certainty to complexity also played out in the humanities. I've already discussed how the two World Wars influenced concepts of cultural

* Often attributed to Margaret Mead, though this has been disputed.

heritage. Much of the militarized librarianship described in chapter 8 was driven by the remnants of the Enlightenment and the belief that true civilization was captured in the European art of writing. Literature was filled with a canon of thought and stories from the Greeks through the Renaissance and the rise of the British Empire—the so-called "classics."

Liberal arts education indoctrinated students in the classics, walking them through how to interpret texts, art, music, and plays. However, just as in physics and operations research, a new paradigm took hold. Modernism, and then postmodernism, and then critical theory, challenged the idea that a reader's, listener's, or viewer's job was to discern the intent of the author. Instead, from texts to architecture, observers glean their own meaning, very much couched in their own experience. From Marshall McLuhan in media studies to Michel Foucault in literary criticism, the work of the creator was to be deconstructed into fundamental elements and understood in a context of power, culture, and history. Knowledge and meaning were not things that could be directly transferred to text and made portable. Rather, the text was a result of knowledge—knowledge that was unique from whatever new knowledge was gleaned by the reader.

Many have seen postmodernism as a shrinking away from scholarly seeking of truth. However, I prefer to see it as embracing the fact that truth is much more complex. For some, this book is about history. For others, it is about data. For still others it is a reflection of a social reality from the privileged position of a white male academic in the United States. The "truth" is that they are all right.

Not everyone agrees with this view. In his 2018 book *Post-Truth*, Lee McIntyre attributes a fair amount of the woes of our current knowledge infrastructure to postmodernism:

> This is not to say that postmodernists are completely at fault for how their ideas have been misused, even while they must accept some responsibility for undermining the idea that facts matter in the assessment of reality, and not foreseeing the damage this could cause.[12]

McIntyre doesn't completely discount the contribution or ideas of postmodernism, but he does point out the extremes these ideas can lead to. He is not alone. Physical scientists are particularly dubious of how some ideas from relativity and quantum mechanics are treated in the work of the humanities.

I could spend a whole book on how science has evolved from certainty and universalism to complexity and probability but, well, that would be another book. Just to state the obvious here, this work is very much influenced by all of these new conceptions of how one comes to know the world. They will be fundamental in building a new agenda for the knowledge infrastructure that veers away from determinism and a belief that the world is a series of information flows that can be tapped through data analysis. Also, to state the obvious, the sciences discussed here were also driven by war and conflict. I would argue, however, that these theories were developed with many steps to validate the ideas outside of funding or direct application to military tasks.

There is one more scientific concept that I have to cover in a book about a knowledge infrastructure: constructivism.

FROM ONE-ROOM SCHOOLHOUSE TO FINDING ONE'S OWN REALITY

If I am examining the knowledge infrastructure as the way we come to know the world, it makes sense to utilize theories and concepts from learning. Just as with the physical sciences, social sciences, and humanities, education and the science of learning have evolved throughout the twentieth and the twenty-first centuries in a familiar pattern.

It wasn't until 1920 that all U.S. states required children ages 8 to 14 to attend school. Even then, school experiences varied widely from state to state. Students in rural areas, for example, had shorter school years to accommodate farm duties,[13] and the one-room schoolhouse was still a fixture of many communities. In more differentiated programs, one with distinct grades and levels, industrialization, mass production, and of course, segregation shaped the classroom. In 1913, the year before the outbreak of World War I, Henry Ford famously introduced the assembly line into car manufacturing. The efficiencies of mass production, particularly when they were so successful in war mobilization, were widely adopted into the classroom as newly formed school districts had to teach thousands of children.

The curriculum of the time was based on memorization and focused on basic math and language skills. The model of education was that students were empty buckets, and teachers would lob bricks of knowledge from the front of the classroom to fill the buckets. Naturally, some buckets were bigger than others. In the words of educational reformer John Dewey in his criticism of formal education, "The child is [seen as] simply the immature being who is to be matured; he is the superficial being who is to be deepened."[14]

Dewey would be part of an education reform movement that introduced concepts such as school being a social institution linked into the community, and its role was to make the student a member of that community. This social function was as important as learning to add and subtract. Dewey argued for instruction that linked to real experience, what today we would call scaffolding and authentic instruction. Other education and learning reformers, such as Jean Piaget, introduced concepts that centered students and their developmental staging in the education process.

Over the century formal education systems, like public schools, have had to find an often-uncomfortable balance between the ideals of learning theory and the practicalities of universal education. Classes are too crowded, with class size determined by budgets over pedagogy. Segregation of Black students into separate and unequal schools was first legal, and then replaced by de facto segregation that exists to this day. Primary and secondary schools have been called upon to do more and more in socializing our children. A curriculum of math and literacy was soon joined by civics, financial literacy, computing, language instruction, and home economics. Schools started as places of learning, and then added missions in health care through the school nurse, mental well-being through counseling, and even food assistance through free and reduced lunch programs.

However, the development of learning theories has not been stopped or constrained by the physical classroom. Learning and how we acquire knowledge is so fundamental to the human condition, learning theory and philosophy have often intertwined. John Dewey, for example, was not a classroom teacher, but a philosopher and the father of pragmatism. Pragmatists see knowledge and personal agency as intertwined. That is, you cannot know the world without seeking to affect it. The concept of the knowledge infrastructure as the mechanism to find meaning *and* power in the world is a "pragmatic" idea.

More recently, the idea that learning is grounded in the individual has been developed under the rubric of constructivism. The basic tenets of constructivism are that:

- Knowledge is constructed, rather than innate, or passively absorbed
- Learning is an active process
- All knowledge is socially constructed
- All knowledge is personal
- Learning exists in the mind[15]

There are different flavors of constructivism. For example, social constructivism states that meaning is held collectively in society versus in the individual. Radical constructivism does away with a shared reality and defines truth and reality as unique to an individual's experience. You can see that constructivism and postmodernism have a lot in common.

The philosophical grounding of constructivism has been expanded to include cognitive studies and even neurology to explain how our mind and the physical brain learns. Cognitive studies show us that we are constantly engaged in an internal dialog known as metacognition. So, when you are reading this text, you are constantly deciding if you agree with what I am saying, relating it to what you already believe, or even framing counterarguments. In many training and learning environments, this is referred to as critical thinking.

FROM DECONSTRUCTION TO CONSTRUCTING THE KNOWLEDGE INFRASTRUCTURE

With all of these different theories and ideas, I can now describe the knowledge infrastructure as a complex system of agents. Agents like reporters, librarians, and professors, and agents like technology companies and media conglomerates. The knowledge infrastructure is an open system, meaning it doesn't have hard boundaries, so it can also include friends and neighbors. We use it to construct knowledge, and in doing so seek to change the infrastructure to our needs. We are also shaped by the infrastructure because it can either increase or decrease the diversity of narratives, facts, and ideas that we can incorporate into our worldview. We use the knowledge infrastructure to reduce personal uncertainty and to socially develop common meanings.

The knowledge infrastructure is not a deterministic system of data that can be collected in a neutral way and mined for meaning. Nor is it a simple system in which we consume information or passively provide data. Our actions, and the actions of data collectors, shape the system. Dataism, which focuses only on gathering what is and seeking conclusions, can never produce Laplace's demon because no system can account for all the variables required, nor understand the complexities of a person making do in a world of imperfect information (satisficing). What's more, the very act of creating and acting upon data changes the environment in which those acts are conducted. Having an ad follow you from site to site changes your behavior, which changes the data collected, which changes the way the ad follows you, which again changes your behavior. It is this recursion of learning that dooms determinism in how we make meaning and seek out power.

So, what is the way forward? If massive-scale data processing doesn't provide us with a path to go beyond the world that is to the world we want, what does?

NOTES

1. Aria Bendix, "More Than a Dozen Researchers Predicted How the US's Coronavirus Outbreak Will End. They Estimated Nearly 200,000 People Could Die by the End of the Year," Business Insider, March 26, 2020, https://www.businessinsider.com/coronavirus-deaths-us-predictions-social-distancing-2020-3.

2. Nurith Aizenman and Ayesha Rascoe, "What Does COVID-19 Modeling Show, and How Can U.S. Lessen the Pain?" April 1, 2020 *Morning Edition*, produced by NPR, Podcast, MP3 audio, 8:31, https://www.npr.org/2020/04/01/825096375/what-does-covid-19-modeling-show-and-how-can-u-s-lessen-the-pain.

3. Matt Finch, "Matt Finch Real Time" (recorded interview), Librarian.Support, March 19, 2020, https://librarian.support/matt-finch-real-time/.

4. Gerald James Whitrow, "Pierre-Simon, Marquis de Laplace," *Encyclopedia Britannica*, March 19, 2020, https://www.britannica.com/biography/Pierre-Simon-marquis-de-Laplace.

5. Pierre-Simon Laplace, *A Philosophical Essay on Probabilities* (New York, 1902), 4.

6. While I would argue that the principle of falsifiability has always been part of the scientific process, it was formally described by Karl Popper as a way of distinguishing scientific endeavors like physics from other types of inquiry such as astrology and pseudoscience. As a starting point, see "Criterion of Falsifiability," *Encyclopedia Britannica*, February 24, 2016, https://www.britannica.com/topic/criterion-of-falsifiability.

7. T. S. Kuhn, *The Structure of Scientific Revolutions* (Chicago: University of Chicago Press, 1970).

8. William Thomas, “History of OR: Useful history of operations research,” *OR/MS Today* 42, no. 3 (2015), https://www.informs.org/ORMS-Today/Public-Articles/June-Volume-42-Number-3/History-of-OR-Useful-history-of-operations-research.

9. Devan Cole, Kevin Bohn, and Dana Bash, “US Could See Millions of Coronavirus Cases and 100,000 or more deaths, Fauci Says,” CNN, last modified March 30, 2020, https://www.cnn.com/2020/03/29/politics/coronavirus-deaths-cases-anthony-fauci-cnntv/index.html.

10. M. Mitchell Waldrop, *Complexity: The Emerging Science at the Edge of Order and Chaos* (New York: Touchstone, 1992).

11. John Henry Holland, *Hidden Order: How Adaptation Builds Complexity* (New York: Addison Wesley, 1995).

12. Lee McIntyre, *Post-Truth* (Cambridge, MA: MIT Press, 2018), 127.

13. “Teacher’s Guide Primary Resource Set: Children’s Lives at the Turn of the Twentieth Century,” Library of Congress, accessed July 29, 2020, http://www.loc.gov/teachers/classroommaterials/primarysourcesets/childrens-lives/pdf/teacher_guide.pdf.

14. John Dewey, *The Child and the Curriculum* (Chicago: University of Chicago Press, 1902).

15. Saul McLeod, “Constructivism as a Theory for Teaching and Learning,” Simply Psychology, accessed July 29, 2020, https://www.simplypsychology.org/constructivism.html.

12

Pandemic

From the Spanish Flu to COVID-19

By the time you are reading these words, I can only hope that the global COVID-19 pandemic has passed. I can only hope that a second, more virulent wave has not occurred. I can only hope that the we are still a society in large part intact. If we are, however, it is not because we learned our lesson in 1918.

FROM THE PHILADELPHIA PARADE TO THE PUBLIC SPHERE

The great influenza of 1918 has become a stark and relevant case study for examining the knowledge infrastructure and the role of data and media in our daily lives with the COVID-19 pandemic, a novel coronavirus that has killed hundreds of thousands and infected millions worldwide. The influenza pandemic of 1918 has a strong connection to the past century of conflict and warfare. It, too, was shaped by World War I. Yet to examine it, I will look not to the battlefields of Europe, but instead begin in Haskell County, Kansas, population 1,720.

> It is impossible to prove that someone from Haskell County, Kansas, carried the influenza virus to Camp Funston. But the circumstantial evidence is strong. In the last week of February 1918, Dean Nilson, Ernest Elliot, John Bottom, and probably several others unnamed by the local paper traveled from Haskell, where "severe influenza" was raging, to Funston. They probably arrived between February 28 and March 2, and the camp hospital first began receiving soldiers with influenza on March 4. This timing precisely fits the incubation

period of influenza. Within three weeks eleven hundred troops at Funston were sick enough to require hospitalization.[1]

From Funston, newly conscripted soldiers spread the flu to nearly every other Army base in the United States and from there to England and then the front lines across Europe. The flu itself spread in two waves. The first wave was considered mild. While it infected a large number of people around the world, it killed relatively few—no more than a standard flu season. However, a second wave of infections from August 1918 was lethal. The War to End All Wars killed 20 million people. The flu killed 50 million and infected 500 million—28 percent of the global population at the time,[2]

The 1918 flu was known as the Spanish flu. The name did not come from the place the flu originated. By the time the flu hit Spain, it was not only widespread around the world, but had already played a role in changing the outcome of battles along World War I's Western Front. No, it was known as the Spanish flu because of Spain's free press.

Spain was a neutral country in World War I. It did not establish censors or clamp down on the newspapers. So, when flu infections spiked around the country, the media reported it. In the warring countries, the free press had either been suppressed through outright censorship or distorted by propaganda efforts that downplayed bad news. The suppression of vital influenza information in the midst of the outbreak fueled the spread of the disease and led directly to greater number of dead.

In John Barry's book, *The Great Influenza*, which I simply cannot recommend more highly,* he tells the story of a "Liberty Loan" parade in Philadelphia in September 1918, meant to sell millions of dollars of war bonds. By the time the parade was held, the flu was tearing through naval bases and shipyards in the city. Barry tells of exhausted nurses putting toe tags on sick soldiers to save time—death was inevitable. Yet the political powers of the city, in keeping with the active U.S. propaganda campaign that suppressed all negative news (and inflamed racial hate), insisted the massive gathering go on as scheduled. Two days later the flu swept through the overcrowded city like fire through dried timber, making Philadelphia the hardest-hit city

* Seriously, one of my favorite books of all time and well before the current pandemic. While it covers the spread of the flu, *The Great Influenza* is as much a story about the development of the medical field and introduction of science into the field.

in the country: "Within 72 hours of the parade, every bed in Philadelphia's 31 hospitals was filled. In the week ending October 5, some 2,600 people in Philadelphia had died from the flu or its complications. A week later, that number rose to more than 4,500."[3]

To be clear, we live in fundamentally different times, and the COVID-19 pandemic is not the Spanish flu. For one, our health care system is in much better shape. Barry's book is as much a story of the rise of modern medicine as it is about the flu. In the past century, the nation's and the world's knowledge infrastructure was a fundamental part of growing and then incorporating modern virology and medical education. Scientific journals link scholars and their work to databases and clearinghouses discussed in chapter 6 to fuel research and medical education. The advances in medicine not only spurred an increase in average lifespan, it also fostered great trust in medical professionals, as seen in Chapter 10.

We are also now in fundamentally different times because of the evolution and greater sophistication of propaganda, censorship, and the media landscape. We see in the recent coronavirus pandemic the same problems play out in educating the public about the virus response as we did in 1918. In 1918, rumors were widely shared that the German "Huns" had spread the virus. Today? It's the Chinese,[4] with overt racist overtones in calling the novel coronavirus the Chinese virus, or the much more hateful "Kung Flu."

Philadelphia politicians in 1918 worried about the economic impact and political backlash, at first ignoring the advice (and warnings) of the medical community, then centralized messaging in the political system and then, finally, began to act. In the current pandemic, President Trump dismissed the potential spread of the coronavirus in December 2019. In February he channeled all federal communications on the virus through his vice president. In March Congress appropriated $8.3 billion to fight the virus ($6 billion more than the administration initially asked for). By March 13, the president declared a national emergency.[5] Different actors—same script.

Before we dive deeper into how the coronavirus pandemic highlights faults in the knowledge infrastructure, please understand that these failures are not just those of President Trump or specific to this particular outbreak. The same pattern can be seen in the disastrous hurricane relief efforts in New Orleans and Puerto Rico. Large parts of our knowledge infrastructure are designed around those with power seeking to control the message, the media amplifying those messages to increase audience engagement, and an untrusting fringe seeing all messaging as a plot. A weaponized knowledge infrastructure is poised to always shout but rarely deliberate, because one leads to increased revenue and one leads to complexity that doesn't help the revenue part of the equation. In the mass media, in narrowcasting, even in academic publication—if it bleeds, it leads.

To bring a sense of focus to an unfolding disaster and to use this focus to build an agenda for change, I return to the components of the infrastructure: people, technology, sources, and policy. Through these lenses I will use current (well, current when I wrote them) examples to highlight fault lines and discuss how we can rebalance, repair, and reinvigorate a knowledge infrastructure centered on the human experience and expertise.

I've tried to write each chapter so that as the pandemic recedes in time, the learning and change needed can remain relevant. Of course, only time will tell if I succeed.

NOTES

1. John M. Barry, *The Great Influenza: The Story of the Deadliest Pandemic in History* (New York: Penguin Publishing Group, 2004).

2. Eric Durr, "Worldwide Flu Outbreak Killed 45,000 American Soldiers During World War I," U.S. Army, August 31, 2018, https://www.army.mil/article/210420/worldwide_flu_outbreak_killed_45000_american_soldiers_during_world_war_i.

3. Kenneth C. Davis, "Philadelphia Threw a WWI Parade That Gave Thousands of Onlookers the Flu," *Smithsonian Magazine*, September 21, 2018, https://www.smithsonianmag.com/history/philadelphia-threw-wwi-parade-gave-thousands-onlookers-flu-180970372.

4. Unless you are Chinese, in which case it's the United States; see Ryan Pickrell, "Chinese Foreign Ministry Spokesman Pushes Coronavirus Conspiracy Theory that the US Army 'Brought the Epidemic to Wuhan,'" Business Insider, March 14, 2020, https://www.businessinsider.com/chinese-official-says-us-army-maybe-brought-coronavirus-to-wuhan-2020-3.

5. You can see the whole timeline here: Ken Dilanian, Didi Martinez, Merritt Enright, Phil McCausland, and Robin Muccari, "Timeline: Trump Administration's Response to Coronavirus," *NBC News*, March 17, 2020, https://www.nbcnews.com/politics/donald-trump/timeline-trump-administration-s-response-coronavirus-n1162206.

13

People

From Empty Campuses to Open Arms

The speed of change in the United States brought on by the novel coronavirus was breathtaking. The first recorded case in the United States was on January 21, 2020. The first death was reported on February 29. Twelve days later President Trump, in an Oval Office address, announced a travel ban for those seeking to enter the United States from Europe (Chinese visitors had already been banned). Within 20 days of the travel ban all U.S. citizens were called upon to practice social distancing. In state after state gatherings of more than 50, then 10, then 3 were banned. By April 2, "at least 297 million people in at least 38 states, 48 counties, 14 cities, the District of Columbia, and Puerto Rico"[1] were being ordered to stay home.

So far in this book we have discussed a number of key knowledge professions: journalists, teachers, professors, librarians, health workers, and government agencies. To some extent all of these people were ordered out of their offices and back to their homes. Journalists started reporting from their homes using Skype and Zoom. Teachers were in their homes delivering, where they could, online instruction to K–12 students. Universities sent their students home and in a matter of a week moved all classes online or canceled them where they could not. Public libraries closed their doors, left the Wi-Fi on for people to use in the parking lot, and sent the staff home. Nonessential government workers were either sent home or furloughed.

In many ways it is remarkable to see how many of these functions—from news to instruction—were able to quickly adapt. However, the pandemic also demonstrated massive gaps in these systems. When colleges went online, they found out that closing dormitories made vulnerable students homeless. Teachers at K–12 schools quickly discovered they could provide online lessons, but a huge number of their students had no online access—or access to food, for that matter. News organizations that had grown reliant on freelance writers and contributors lost huge swaths of this contingent workforce as revenue dried up in the ensuing economic shock to the country.

However, it should be noted that these knowledge workers were in some ways the best prepared for sheltering in place. They were, by and large, prepared to use digital tools. Their work was often location-independent—they were able to write, lecture, and operate wherever they had a laptop and a network connection. Their functions were also seen as vital. Education for the next generation had to continue. Sharing information had to continue.

Health workers, on the other hand, experienced a much more complex relationship between technology and physicality. While some functions could be conducted via telehealth, many treatments required being at a physical facility. These workers not only suffered with exposure to the virus, but a lack of vital medical equipment. Huge numbers of non-frontline employees in health systems were also furloughed as high-revenue elective procedures were canceled.

However, for this examination of the knowledge infrastructure's people component, I want to focus on a very particular issue with people and health information—the reduction of public-sector health expertise. In the first three years of the Trump administration there was a coordinated effort to reduce scientific, medical, and domain expertise across the government. This was most notable in foreign affairs, with steady cuts to the diplomatic services at the U.S. Department of State. However, there were also many cuts in the field of health care and medicine.[2]

FROM AIDS TO CLIMATE CHANGE

Prior to the pandemic, the Trump administration shut down the global health security unit of the National Security Council. It eliminated a $30 million "complex crisis" fund. It reduced federal spending on national health by $15 billion, and even as the outbreak was spreading, the administration argued for a 16 percent cut to the Centers for Disease Control and Prevention.[3] All of these cuts defied the need for pandemic preparedness, which was a priority for previous administrations combatting the spread of AIDS, swine flu, SARS, avian flu, and Ebola.

This decrease in investment and preparedness for health emergencies happened at local and state government as well.

> In 2000, the Institute of Medicine issued a report titled "Public Health Systems and Emerging Infections: Assessing the Capabilities of the Public and Private Sectors." One of the report's findings was that the basic infrastructure of the American public-health system, particularly at the state and local levels, is eroding. With that deterioration comes a diminished capacity to predict, detect, and respond to an emerging infectious disease. There have been scores of similar reports issued in the past twenty years—nearly each of which, like this one, was essentially ignored. This report was edited by the late Nobel laureate Joshua Lederberg, whose insights into the role of viruses and bacteria in the life of this planet should be read by every elected official with the power to appropriate even a single dollar toward our health care.[4]

Clearly no government has unlimited resources to have experts on staff ready for any and all potential emergencies. We all hope a global pandemic is a very rare occurrence, and that hindsight is 20/20. However, this exact contingency was planned for and justified within the past 10 years. In 2016 Obama and Trump cabinet-level officials even participated in a joint briefing on the possibilities of global pandemics just like COVID-19.[5]

What's worse is that these cuts represent a worrying trend in regard to the knowledge infrastructure: a systemic devaluation of science and expertise in the public sector. From cuts at the CDC, to the decimation of scientists at the Environmental Protection Agency (EPA), the Trump administration in particular has greatly limited the voice of scientists:

> Budget cuts are only one highly visible strategy. Other executive actions are eroding the capacity of our nation's science agencies. For one thing, Trump officials are taking advantage of additional methods to reduce agency staffing. In the fine print of the president's [2018] budget proposal are reductions in staffing by 20 percent or more in some bureaus (the EPA, for example), often with science

> programs faring the worst. There are buyout offers for eligible employees and staff transfers to shut down specific areas of work. Virtual hiring freezes have been put in place for most civilian agencies. And there are ongoing consultations on how to conduct "reductions in force," otherwise known as layoffs.[6]

This is not just a matter of having enough personnel on staff to respond in an emergency; it is also about the loss of internal advocacy.

As I said, no system, no profession, no person is neutral. We all have biases. This is not a flaw or a shortcoming—it is an asset. It means, as humans, we have passion and an understanding that our words and deeds will have an impact on the world around us. Scientists within the public sector of government provide a point of view pushing cities and society as a whole toward rationalism and shared understandings. Just as falsifiability is a cornerstone of science, so is openness. The reduction in medical and scientific expertise means less advocacy for grounded decisions and transparency.

What is increasingly taking the place of science as a voice is ideology. A study conducted by the Union of Concerned Scientists in 2018 found that:

> scientific experts in the federal government are essential to ensuring that policies are grounded in the best available science. Yet the Trump administration has presided over an unprecedented hollowing out of the federal science capacity. Eighteen months into his administration, President Trump had filled just 25 of 83 posts designated by the National Academy of Sciences as "scientist appointees"—far fewer than in previous administrations. And science-based federal agencies are losing expertise as scientists leave government service through early retirements, hiring freezes, and other workforce reduction measures. Staffing at the EPA, for example, is at its lowest level in 20 years—and the Trump budget calls for a further reduction.[7]

This reduction has a larger effect on other knowledge workers. As discussed in chapter 6, federal research dollars have become an integral part of how research universities are funded. Federal agencies design the research grant programs available, select which scientists get funding, and can strongly influence the type of questions scientists pursue. So, ideology can influence work done in and outside of a government agency.

Once again, we've seen this throughout history. From the atom bomb to operations research to the internet to encryption, the government has been a large determinant of the technologies, policies, sources, and experts that are available through the knowledge infrastructure and can be mobilized. These efforts have been driven by war funding. When the government scales back on science, it increasingly pushes scientists to the private sector where consumer demands push an agenda. A clear example of this was the politically driven 25-plus-year moratorium on gun violence research.[8]

So in March 2020, we found ourselves with limited voice, capacity, and influence in the scientific community that could be marshalled toward the global pandemic. The messages were being increasingly crafted through the lens of politics, and even those with expertise in reacting to the crisis were now doubted after decades of a devaluation of rationalism over belief, dogma, and ideology. The same thing is happening with regard to climate change, but over decades instead of weeks. Add to that list smoking, leaded gas, housing, and civil rights. Whether it is the 1918 call for unity around the war winning out over good science, or today's concern for the economy over human life, who is participating in conversations before as well as during a crisis matters.

FROM AMSTERDAM TO CHARLOTTESVILLE

I will be honest; I have a bias toward science. Is this bias any more justified than, say, a bias toward a given political ideology or even toward a knowledge system like religion? And if so, why? If not, then are there no negative biases? Racism, for example?

To me, the answer lies in the concepts of diversity and inclusion. Not the caricature of diversity that is often presented—as a politically correct purity test that avoids hurt feelings or preferences one group for appearances' sake. I am talking about diversity as mutual respect and as seeing value in alternative views based upon different life experiences. Put simply, the richest knowledge comes from the richest and most diverse set of inputs. Put aside for a moment issues of class, religion, race, and perceived disability (just for a second). Take Kuhn's look at science as a series of paradigms. In science Kuhn saw groups getting stuck—unable to ask important questions or answer new problems—because their frame of reference became fixed. Science requires people new to a domain to inject new questions and energy. It is also the limitation of a dataist approach—data at its best can act as a constant mirror to what is (and can be) recorded, but not a as doorway to what might be.

In chapter 9 I discussed the importance of diverse narratives. People need to see a spectrum of stories to not only better understand a culture, but to envision their own possibilities. Why is it so important for girls to go to school in Afghanistan? Because they are exposed to stories of what they can aspire to, and so they have the power to share their knowledge, culture, and experience to enrich the world.

In his outstanding book on the history of public libraries in the United States, Wayne Wiegand chronicles the fight to get the "literary novel"—what we simply refer to as a novel today—banned from libraries at the end of the nineteenth century. He quotes an 1882 newspaper article on who reads novels: "Schoolchildren; factory and shop girls; men who tended bar, drove carriages, and worked on farms and boats; and finally, fallen women, and, in general, the denizens of the midnight world, night-owls, prowlers, and those who live upon sin and its wages."[9]

Critics would write of how the novel could push girls to think above their station and give farm boys dreams of the wild west instead of what they should be doing . . . toiling in the fields. The sharing of diverse ideas and experiences enhances society because it opens up possibilities. What is true in science and literature is certainly true for class, religion, and race.

The idea of diversity and inclusion appeals to society's loftiest ethical and moral goals. But it is also immensely practical. If we were to study the knowledge infrastructure well before World War I, say in Muslim Spain in the eighth and ninth centuries, we would find cosmopolitan cities that welcomed all religions and races and therefore thrived. The Iberian Peninsula in the Middle Ages revolutionized architecture, mathematics, and government and was a direct precursor to the Renaissance in Italy.

The same story could be told of Amsterdam 500 years ago: "In becoming the melting pot of Europe in the 1500s and 1600s, the city set the template for modern urban life."[10] In the seventeenth century, Amsterdam became the wealthiest city in the western world. Yes, it was a colonial power, but it also accepted refugees and became a city that welcomed all religions and races. It used diversity to attract oppressed skilled workers and merchants driven from other countries by hatred and bigotry.

Diversity of people, of backgrounds, of skills, of perspectives are essential in a well-functioning knowledge infrastructure, but only if the differences are respected and the biases used to further learning. Learning is about seeking out new understandings and, ultimately, agreements among people—agreements like new theories in physics, new trade agreements, and new social norms. This only happens when people are open to learning.

Racism and prejudice, while types of bias, are ultimately a state of willful ignorance. They deliberately deny the positive effects of diversity and inclusion. Purposefully living in a state of ignorance is counter to what a just and growing society seeks out. Therefore, giving voice to racism does not further the conversation or learning of a community.

Now, stating that racism is bad is a pretty obvious call. Talking about not promoting racists views is also a pretty easy call. Here's the thing: stating as a journalist or librarian or professor that we must choose to actively counter racism, even to the extent of not giving racists a platform to speak, is another thing altogether. It means acknowledging that, as knowledge professionals, they are not neutral, and that as professionals they must take a stand against part of their own communities. And that is hard.

Shouldn't the knowledge infrastructure be a place for all voices in the society? No. The knowledge infrastructure is not a neutral microphone placed in a town square, open to all comers. It is a platform of learning that acknowledges the full range of views in a community, and with the community develops and supports a learning narrative that pushes against racism and bigotry.

Society cannot ignore the systemic racism and the passive attitude that allows white supremacy, anti-Semitism, misogyny, or religious intolerance to fester. Understand also that fighting the ignorance of racism and intolerance doesn't start nor end at the door of a library, college, or what airs on the 6 o'clock news. It means fighting for better transportation options to allow greater economic opportunities for underserved people. It means not only reporting on housing discrepancies, but lobbying to eliminate them.* It means believing those who are marginalized when they tell you something is wrong, inequitable, or abusive, and doing something about it. The knowledge infrastructure is ultimately a motor for societal change.

---

* A great example of action with a foundation in science is the Eviction Lab at Princeton University: https://evictionlab.org. Not only does the lab collect and analyze data; it seeks to expose and end a housing crisis situated in historical inequity and racism. It is a project that demonstrates the difference between data science and dataism.

FROM HEARING AIDS TO TRANSIT MAPS

Perhaps the greatest way that the knowledge infrastructure can produce change is by acknowledging how our concepts of diversity and the traits we consider normal are in fact socially constructed. Take, for example, the concept of disabilities.

According to a 2018 CDC information sheet, one in four Americans has a disability; two in five adults over age 65 have a disability.[11] But what does that mean . . . once again, what's behind the data? Dr. Clayton Copeland, a professor at the University of South Carolina (and, I am proud to say, a colleague of mine), points out that what constitutes a disability is not based on some objective definition.

For example, we might all say that an inability to hear (deafness) or to see (blindness) is a disability. Yet do we consider people who wear glasses or use a hearing aid later in life handicapped? Autism is a disability, but what about depression or post-traumatic stress disorder? Certainly, we might say they are debilitating, but are the people who experience them labeled as disabled? I, for example, have a compromised immune system from a bone marrow transplant to treat cancer. Am I disabled? Was I disabled when I had cancer?

We don't normally think of people as disabled if there is an accommodation that can allow for "normal" everyday operation: glasses, hearing aids, medication like insulin for diabetes. Yet there is nothing inherent in the conditions that separate these accommodations from wheelchairs, ventilators, or screen readers needed by some people.

People like Dr. Copeland who both study and advocate for what has traditionally been labeled a disability are seeking to change the narrative and the terminology that reinforces certain types of diversity as negative. The use of "differently abled" or "specially abled" may seem the height of political correctness, but it's also a way of showing respect for another person. It also acknowledges that differences can also lead to new abilities and new means of development.

A central concept in the equity and inclusion community is universal design. Yes, I know I spent a whole chapter on how nothing can truly be universal, but that doesn't mean there aren't times to try. Universal design seeks changes in systems (physical buildings, software, policies, laws) that will benefit everyone, and thereby enrich the diversity of people and thoughts in the knowledge infrastructure.

There are some classic examples of universal design in action. To accommodate people with mobility issues, sidewalks are now constructed with small ramps at intersections. It turns out the majority of users of these ramps are parents and caregivers with strollers. Also, from experience, travelers pulling suitcases through cities. Accommodations originally intended for some, benefit many.

Public institutions (schools, libraries, government buildings) wanting to better serve people on the autism spectrum will put video tours of their spaces online. For many people with special sensory needs, familiarization with an environment reduces stress. Seeing a place before they have to encounter it for the first time helps. This also helps people with Alzheimer's disease and other memory conditions.

In March 2018 Google introduced "wheelchair accessible" transit maps that help not only people in wheelchairs, but anyone with mobility issues—like the elderly, parents with strollers, businesspeople with bags, and cancer patients exhausted from chemotherapy. There are other examples where technological developments first applied to commerce or conflict are now helping to provide equity online. Machine learning and artificial intelligence are automatically creating transcripts on YouTube—making video content accessible for people using screen readers and those with hearing impairments alike. The same technology that powers augmented reality—placing virtual objects in a physical world—allows enhanced vision through cities and buildings to people with visual impairments. Even cell phones and Bluetooth headphones originally designed for listening to music are being approved for hearing enhancement.

These actions—video maps, transit maps, sidewalk ramps—are more than just niceties. They are a recognition of the inherent value of people, regardless of this match to a shifting norm. A person's value to society should never be measured against an average. These decisions help people, but attention to diversity can also save lives.

In the 1,000-year flood in Columbia, South Carolina, in 2015, rescue workers set out on boats to save people cut off by rising water. One house on a rise had become an island surrounded by the rising floodwaters, now contaminated with runoff from the local sewage treatment plant. When the rescue workers arrived, they encountered a family unable to leave; a man in the house was in a life-sustaining wheelchair. Unable to accommodate the chair in the boat, the rescue workers had to leave and hope the floodwaters crested soon. Thankfully, the waters did indeed stop rising, sparing the man's life.

In these dire physical conditions, the value of human life becomes acute. Yet our knowledge infrastructure also serves as a vital lifeline to those who have mobility issues, or require medical confinement, or must overcome debilitating mental health conditions. Internet access, telework, online education, and telehealth are all essential services and should be available to all who need them, not simply those who can afford them or who have the "right" capabilities or accommodations.

FROM DEATH WARDS TO MARATHONS

At the outbreak of World War I, cancer was considered not only a lethal condition, but a shameful one. With no effective treatment, victims of lung cancer, breast cancer, childhood leukemia, lymphoma, and myriad other cancers died, hidden away in family homes or hospital wards set up to accommodate the dying. Many saw cancers as the result of moral failings or weakness of character. Through concerted public campaigns and funding of research, first surgical, then chemical, then radiological treatments were developed that extended life and even cure the sick. In the span of a century, society chose to transform cancer from a shameful failing to a heroic cause. Well, in most cases.

The National Institutes of Health (NIH) spends over $5 billion a year on cancer research. Every year millions of dollars are raised to fight breast cancer through marathons; leukemia and lymphoma through Light the Night walks; and colon cancer through the annual Colon Cancer Challenge. Philanthropy for all kinds of cancer is up, except for lung cancer.

Lung cancer is the second most commonly diagnosed cancer and the leading cause of cancer deaths in the United States.[12] Yet,

> it is unclear what accounts for the high rate of nontreatment observed in patients with lung cancer. However, emerging research suggests that negative perceptions of lung cancer, such as blame and hopelessness, may play a role. These negative perceptions may be caused by its association with smoking, the thought that the disease is self-inflicted, and its high mortality rate. Consequently, lung cancer carries the cumulative burdens of social stigma and being a leading cause of death, and these burdens extend to never-smoking lung cancer patients. Studies have also shown that lung cancer stigma is associated with greater patient-reported symptom severity.[3]

To be clear, people are dying of lung cancer, and suffering more, because of stigma and the subsequent discrimination. When you get breast cancer, or lymphoma, or pancreatic cancer, you are a victim, and fate is fickle. But if you get lung cancer, it is because you smoked, and it is your fault. Except, of course, for the 7,330 people who die each year from secondhand smoke, or the 30,000 who get lung cancer with no connection to smoking whatsoever.

Addiction, depression, and AIDS were all seen once as shameful conditions aligned with personal failings (and in the case of AIDS, exacerbated by homophobia). Here again we see the vital need for a functioning knowledge infrastructure grounded in the value of human life and rationalism. People died from shame and ignorance and still do, all at the time backed up by political agendas, prejudice, unchallenged certainty, and, yes, even skewed data analysis. Data drawn from unscientific samples, or lacking context. AIDS was associated with gay men because that's the community first struck by the virus and suffered societal discrimination, not because it was a "gay man's disease." The novel coronavirus is not a Chinese virus just because China had the first hotspot. AIDS and COVID-19 do not discriminate, though people and administrations definitely do.

At the core of the knowledge infrastructure are people and their search for meaning and power. Knowledge is not cold or objective or unbiased. Knowledge is passionate and informed, and culturally competent. It is the result of hard-won experience and an innate human drive to know more, to do more, to be more. We must acknowledge, therefore, that all the components of the infrastructure are also the result of passion, and seek to guide the knowledge infrastructure for the betterment of all, not simply the majority or the "normal" or the owners of the well.

NOTES

1. Sarah Mervosh, Denise Lu, and Vanessa Swales, "See Which States and Cities Have Told Residents to Stay at Home," *New York Times*, last modified April 20, 2020, https://www.nytimes.com/interactive/2020/us/coronavirus-stay-at-home-order.html.

2. Sonam Sheth and Gina Heeb, "Trump Spent the Past 2 Years Slashing the Government Agencies Responsible for Handling the Coronavirus Outbreak," Business Insider, February 25, 2020, https://www.businessinsider.com/trump-cuts-programs-responsible-for-fighting-coronavirus-2020-2.

3. Joseph Zeballos-Roig, "Trump Defends Huge Cuts to the CDC's Budget by Saying the Government Can Hire More Doctors 'When We Need Them' During Crises," Business Insider, February 27, 2020, https://markets.businessinsider.com/news/stocks/trump-defends-cuts-cdc-budget-federal-government-hire-doctors-coronavirus-2020-2-1028946602.

4. Michael Spector, "The Coronavirus and the Gutting of America's Public-Health System," *New Yorker*, March 17, 2020, https://www.newyorker.com/news/daily-comment/coronavirus-and-the-gutting-of-americas-public-health-system.

5. Nahal Toosi, Daniel Lippman, and Dan Diamond, "Before Trump's Inauguration, a Warning: 'The Worst Influenza Pandemic Since 1918,'" Politico, last modified March 16, 2020, https://www.politico.com/news/2020/03/16/trump-inauguration-warning-scenario-pandemic-132797.

6. Andrew A. Rosenberg and Kathleen Rest, "The Trump Administration's War on Science Agencies Threatens the Nation's Health and Safety," *Scientific American*, January 1, 2018, https://www.scientificamerican.com/article/the-trump-administration-rsquo-s-war-on-science-agencies-threatens-the-nation-rsquo-s-health-and-safety.

7. "Science Under Trump: Voices of Scientists Across 16 Federal Agencies," Union of Concerned Scientists, August 7, 2018, https://www.ucsusa.org/resources/science-under-trump.

8. Nidhi Subbaraman, "United States to Fund Gun-Violence Research After 20-Year Freeze," *Nature*, December 17, 2020, https://www.nature.com/articles/d41586-019-03882-w.

9. Wayne A. Wiegand, *Part of Our Lives: A People's History of the American Public Library* (New York: Oxford University Press, 2015), 60.

10. Russell Shorto, "Life in Amsterdam: The Benefits of Openness," *AMS Magazine*, May 16, 2016, p. 23, https://issuu.com/iamsterdam/docs/ams_regular_lr/24.

11. "Disability Impacts All of Us," Centers for Disease Control and Prevention, last modified September 9, 2019, https://www.cdc.gov/ncbddd/disabilityandhealth/infographic-disability-impacts-all.html.

12. N. Sriram, Jennifer Mills, Edward Lang, Holli K. Dickson, Heidi A. Hamann, Brian Nosek, and Joan Schiller, "Attitudes and Stereotypes in Lung Cancer versus Breast Cancer," *PLOS ONE* 10, no. 12 (2015), https://doi.org/10.1371/journal.pone.0145715.

13. Ibid.

14

Technology

From Zoom Bombs to the Shin Bet

The COVID-19 pandemic has demonstrated the new central role technology has in the knowledge infrastructure in the United States and across the globe. In a matter of one month in early 2020, life in the United States changed. Classes were attended online, groceries were ordered online, movies skipped the theater to stream online. To be clear, not all Americans did all of these things online; the pandemic demonstrated that the digital divide was a chasm that left too many of our most vulnerable citizens behind. I will get to that in chapter 16. But for now, I will focus on the fractures in the technology aspects of the knowledge infrastructure.

The global pandemic did not change the fact that Americans study, shop, and watch movies online in a system built around a surveillance business model that trades personal data and privacy for service. It only forced more people to do more in that system.

Take the explosion in the use of Zoom, a video conferencing solution. In March 2020 Zoom was so widely used, it generated its own set of memes and jokes. There were Zoom BINGO cards for which participants get points for "forgetting to turn off embarrassing text notifications while screen sharing," or "random child in the background." There were Zoom parties. And then there was Zoom bombing.

Zoom bombing* is when inadequate security on the part of the person organizing a video conference allows uninvited users to join and disrupt a meeting. It can be anything from a prankster logging on, yelling, and logging off to uninvited users entering a Zoom session and sharing pornography.[1] As many college and university instructors hastily adopted Zoom for classes forced online, they found pranks could become disruption that could become online harassment:

> "I didn't notice it until a student on chat said something about it," said Gharavi, an associate professor in ASU's [Arizona State University's] School of Film, Dance and Theater. Participants were using fake screen names, some of which he said were very offensive. "The chat window became incredibly active. Most of the comments were not on topic. They were vulgar, racist, misogynistic toilet humor. I would barely even call it humor."
>
> Gharavi was not alone. The University of Southern California reported similar incidents occurring while professors taught classes on the same platform, indicating that the massive migration of college classes online due to the public health crisis came with a new threat—one that's technical rather than biological. The professors were the victims of "Zoombombing"—the "Zoom" in this case being the online meeting and course-hosting platform, and the "bombs" typically taking the form of racist vitriol or pornographic content shared with the group by an unwelcome user.[2]

Unplanned and rushed online instruction took on a more menacing tone as people on the more extreme ends of the political ideological spectrum (left and right) logged in to large classes and recorded what they felt was evidence of ideological bias on the part of professors:

> The coronavirus-prompted shift to remote teaching was stressful enough for faculty members before Charlie Kirk weaponized online learning. On Sunday the founder of the conservative political-action group Turning Point USA told college students whose professors had switched to online classes to share with Turning Point videos of "blatant indoctrination."
>
> "Now is the time to document & expose the radicalism that has been infecting our schools," he tweeted. "Transparency!"[3]

---

* This phenomenon is happening so fast, the name has not yet settled. I've seen it as Zoom bombing, Zoombombing, and Zoom-bombing.

This created a chilling atmosphere and an unwillingness on the part of some professors to engage in conversations they may have had in a physical classroom.

Zoom bombing is an example of aspects of the knowledge infrastructure, such as security and encryption, that are either ignored or not set as the default. Zoom, and other video conferencing applications, can be made secure. However, doing this requires more training and awareness than a professor trying to move a class online for the first time ever received in a chaotic week of transition. Yet these demonstrations of how unaware people can be regarding data, privacy, and security issues when online pale in comparison with what happened in Israel.

Early on March 17, 2020, the cabinet of Israeli Prime Minister Benjamin Netanyahu had approved the use of cell phone location monitoring to control the spread of the coronavirus.[4] On March 18, some Israeli citizens started getting text messages stating that they may have been exposed to the coronavirus and should self-isolate for 14 days. This was most likely surprising, as the alert system was put together in hours with no public debate.

In China, use of cell phone data was only one of a number of intrusive and obvious means of surveillance. To enforce a mandatory quarantine for the citizens of Wuhan in response to the novel coronavirus, Chinese officials had citizen cell phones directly contact police when the phone, and presumably the phone's owner, wandered from allowed locations. Phone data in South Korea was used to create "a publicly available map from cell phone data that people can use to determine if they have come into contact with someone who has been infected with the novel coronavirus."[5] Taiwan set up GPS fences to prevent quarantined citizens from leaving their homes.

There can be no doubt that these actions saved lives. South Korea, in particular, seems to have minimized the impact of the virus through the extensive use of location data and text messaging that alerted people to nearby infections. My point in raising this is not to argue that privacy take precedence over lives. My point is that the capabilities to do all of these location-based enforcements were already in place years before the pandemic.

In fact, the Shin Bet, Israel's internal security and intelligence service, had been collecting the data that would be used in virus tracking since approximately 2002.[6] Citizens in China, Iran, South Korea, and Taiwan, in a flash, had been made aware that a service they were paying for and thought was for their use, was actually also being paid for with their data for the use of the state.

Perhaps more relevant to this discussion is not how the United States is using location data from cell phones in the crisis, but who is providing it. See if you can catch it in this excerpt from the *Wall Street Journal*:

> The federal government, through the Centers for Disease Control and Prevention, and state and local governments have started to receive analyses about the presence and movement of people in certain areas of geographic interest drawn from cellphone data, people familiar with the matter said. The data comes from the mobile advertising industry rather than cellphone carriers.
>
> The aim is to create a portal for federal, state and local officials that contains geolocation data in what could be as many as 500 cities across the U.S., one of the people said, to help plan the epidemic response.
>
> The data—which is stripped of identifying information like the name of a phone's owner—could help officials learn how coronavirus is spreading around the country and help blunt its advance. It shows which retail establishments, parks and other public spaces are still drawing crowds that could risk accelerating the transmission of the virus, according to people familiar with the matter. In one such case, researchers found that New Yorkers were congregating in large numbers in Brooklyn's Prospect Park and handed that information over to local authorities, one person said. Warning notices have been posted at parks in New York City, but they haven't been closed.[7]

Unlike in Israel, China, South Korea, or Taiwan, it is not the phone companies that are providing location and cell phone data . . . it is advertisers. Don't be fooled—the phone companies absolutely have this data and absolutely have shared this data with the government in the past (and have sold it to third parties). But advertisers have this data ready to use and can sell it, resell it, collect it, and "hand that information over to local authorities," without regulation or additional approval.

It is not just the phone companies or the advertisers that have made this kind of mass surveillance possible. The sharing of location and the monetization of this personal data is a product of the phone manufacturers, advertisers, app developers, network hardware providers, government regulators, and you. It is baked into the same systems used by college and elementary students to continue their studies. It is baked into the same systems bankers and artists and travel agents used to work from home. The same technology used to save lives in the virus crisis was just month earlier doing the exact same thing to make money and shape political messages.

What the pandemic did was to make very clear the current model of the knowledge infrastructure shaped over a century of war. Mobilizing during a pandemic uses the same systems that are used to mobilize for war. And we now see that when mobilizing for war, neither personal privacy nor technology are well understood by the common citizen. Also not well understood are the use of propaganda to control national narratives, the role media plays in indoctrinating the next generation, and the constant presence of profit-making from conflict. These may not be as obvious in "normal times" as a pornographic Zoom bomb in your poli-sci 101 class, but they should be.

FROM CERTIFICATES TO BLOCKCHAINS

In addition to technology to support people interacting and doing work (and making data), technology can be used to ease the management of data itself. What the digital knowledge infrastructure needs is an efficient and easy-to-use way for us to control what data we share online. Rather than creating hundreds of individual accounts with various app developers and web services, each becoming their own repository of our data, we need to own our own data and share what is needed. Just think how ridiculous it would be if you had to set up a separate bank account with your money for every company you bought something from, and if each of these accounts had its own terms of use and separate security measures (or lack of them).

As I will discuss in the next chapter, our current system of personal data and privacy control is deliberately confusing and inadequate. I will propose some regulatory ways of addressing this, but in terms of technology, it is time for some serious technology-led changes.

In chapter 2 I talked about securing the web's Hypertext Transfer Protocol (https, rather than http). This system is based on something called digital certificates. Certificates are digital signatures that a site can transmit to both confirm the authenticity of a site (the Google.com page you loaded is indeed the real Google.com) and the means of encrypting data between you and that server (known as a public key).

Google does not issue a certificate to itself, but uses a trusted third party, such as DigiCert or IdenTrust. Never heard of them? That's because you don't directly interact with them as a user. It's like the difference between your bank and credit reporting services like Experian or Equifax. You can get a loan from a bank, but the bank uses credit reports from these firms to determine if they can trust you in the transaction. You don't hear about these services . . . unless something goes wrong, such as poor security on the part of Equifax, which led to the leaking millions of personal profiles that included social security numbers.[8]

The work of secure certificates and web encryption is invisible to the user (minus the padlock icon you probably see in your address bar, or a warning of a wrong certificate you might get from your browser or even Google). Yet these allow the most sensitive of transactions to occur, including banking, signing legal documents, and consulting with your doctor. There are a number of promising models and technologies that could be used to create secure ways of sharing data with advertisers and service providers in a similar way, but where the control of that data stays with you. Shibboleth[9] is one of the oldest examples on the web. In this system, a home institution (this is mostly used in university contexts, so the university is your "home") holds your personal data and credentials (like your email address, your contact information, and the like). When a system wants to get at part of this data, it asks your home institution, which in turn either asks you or only provides a previously agreed-on set of data.

More recently, Apple implemented something it calls Sign in with Apple. With this system, when a site or app asks you to log in, you can have your iPhone provide just the necessary data while all of your information stays (or so Apple says) on your phone, not with Apple. Apple will even create temporary email addresses to mask your real one to prevent you from getting spammed with unwanted emails.

Apple developed its sign in system as a more private solution compared to similar systems from Facebook, Google, and Amazon. You most likely have run into an app or a website asking you to either create an account, or simply "log in with Google," or Amazon, or Facebook. If you choose this option, you actually log in to Google (or Amazon or Facebook), and then the company will log you into the app or site you want. You may see this as a convenient way to not create another account, but what you are really doing is giving these services even more data than you likely intended.

With these solutions, Facebook, Google, or Amazon hold your credentials. What's more, these companies get more data from the app every time you log in . . . even if that app has nothing to do with Facebook or social media (such as what games you are playing or what events you are adding to a calendar). Using your Facebook login and password to share Words with Friends may be easier for you than creating yet another account, but now Facebook tracks which friends you are playing with, when, and where.

We need to use technology to create trusted repositories for our data that are easy to use and maintain, not located within companies that happen to make our phones or provide a service whose password is easy to remember. Think of it like password managers. The fact that all services require their own passwords has led to a whole suite of tools to capture or generate passwords (with at least one capital letter, a symbol, at least 8 characters, and does not spell out a word—and one you haven't used before). Password managers are quickly being replaced with facial recognition, fingerprint scanners, even biometric identifiers built into watches. Yet because these technologies are being developed by tool makers (Apple, Amazon, Google), they will always be controlled by, and for the ultimate benefit of, those makers. We need a public solution. Better yet, we need a solution that isn't dependent on any one single source.

Self-sovereign identity (SSI) systems are software that put all of your identity information, and potentially your sharable private data, in your hands. Rather than creating hundreds of accounts around the web (to buy books, to play games, to share stories, to bank, to check your health records, to register to vote, etc.), you control it all. Then you choose who gets access to what part of that data. To be frank, searches on the term "self-sovereign identity" will turn up a huge number of results, and all will be dripping with jargon, so let me explain the idea with a few simple analogies.

In his article, "How Blockchain Makes Self-Sovereign Identities Possible," Phillip Windley uses the example of a driver's license in the real world.[10] When you go to a bar and show your license for entry, you did not have to create an account and enter into an end-user agreement with the bar in order to prove you are 21. You store that data in your wallet, under your control. What's more, if the bar needs proof that the ID is valid, it can check with the government bureau that issued the ID. All of this checking and providing data is done in the open, under your control, and in a way that can be trusted. What's more, the trust doesn't just come from one source (the person holding the ID), but from technology (protections embedded in the card such as watermarks, paper quality, format, etc.), from the information sources (the bar may not trust you, but what about the agency that provided the ID or provides a check on the credential), policy (not only is underage drinking illegal, but both the ID holder and the bar can be penalized), and the people involved (training, ethics, honesty)—all the parts of the knowledge infrastructure.

There are plenty of other physical world examples showing how you control access to your data. Credit cards, for example. Once you establish a relationship with a bank, you can get a credit card and a series of identity-proving numbers (the credit card number, the expiration date, the three-digit security code on the back of the card). Now you can go to any place that accepts credit cards and use an identity you hold, to get what you want—buy gas, see a movie, subscribe to cable—all with a single credit card in your control. The bank is involved too, mind you. In real time based on those identifying numbers, the card assures a company you're are good for the money. It can even, in real time, ask for additional data to be even more sure of your true identity (a signature, a zip code, the name as it is written on the card). Once again, you didn't have to get a credit card from every gas station, movie theater, or cable provider. Why do we accept having to create privately held banks of our data at hundreds of web sites and apps, when we would never do so in the physical world?

Recent technologies are making self-sovereign identities a reality and showing how these can be more secure as well as more convenient. Namely, Blockchain.

You may remember the discussion of peer-to-peer design and Napster back in chapter 5. Peer-to-peer allows technologies on the internet to communicate directly to each other without a central server. While Napster has gone away, peer-to-peer has not. There is currently a sort of gold rush around using peer-to-peer technologies to create a new form of currency: Bitcoin. Bitcoin is only one example of a larger concept called Blockchain.[11]

Blockchain is a distributed ledger system. Think of it as keeping a log of a certain transactions, like when you paid $5 for a book. When you keep that log, it has a sequence. For example, you paid the money first, then received the book. Keeping this log is important, because if the bookseller claims you never paid for the book, you want documentation. But how can the bookseller trust your records? The solution: you keep a set of records, and the seller keeps an identical set, and maybe a third party also keeps a set. If there is ever a question, all the records can be compared. With technology, we can make this a bit more secure. Now the records kept by the three parties will include an encrypted code based on the transaction data. If someone tries to alter the records, the encrypted code will no longer match up to the information.

With Blockchain, these records are called ledgers, and instead of three parties, there are thousands and thousands of them. Every time a transaction occurs (stored in a "block"), it is added to all of the ledgers. The "chain" is that every block, on every ledger, is sequenced. The money was exchanged, one block, then the book was delivered, next block. If the seller wants to claim he or she wasn't paid, his recollection would have to match thousands and thousands of ledgers stored all over the internet. How does this help with privacy?

While you store the data you want to share, how can the third person rely on it? How does this website know you are 21 years old, or that you are, well, you? All of this data can be stored as encrypted information in a public (or private) ledger system and checked for accuracy and to assure that no information has been altered. So if the state says "this is your digital driver's license" by adding a block to your chain, you can't change it without messing up the record. You get the trust that comes from state authority, the convenience and security of managing your own data, and organizations still get the data they need to operate . . . but only the data they need and that you are willing to grant.

Now when Google wants information, it has to get it from you. When Google adds information to your records, it can do so with authority—just like the government adding information on whether you can drive or drink. Are there still security concerns with self-sovereign systems, such as hacking? Absolutely. But now that security is under your control (or more likely, the control of a trusted party), and not in the hands of hundreds (or thousands) of apps and companies storing their own data stores. You may trust Google to secure your email, but Candy Crush? And Farmville? You may trust Google to secure your data, but do you trust it to only use that data in a way you are comfortable with?

FROM AI TO IA

Such a system can begin to bring the use of artificial intelligence in line with societal norms and agreements. There is no doubt that using deep learning techniques to improve cancer diagnosis is a good thing. No doubt many people would give permission to opt in to that work. That's a very different default than "we have your data and want blanket permission to use it." The value of data becomes apparent to citizens, not just the aggregators that collect and resell it.

Part of opting in, that is, part of having to consciously choose, audit, and retain control of data, is the ability to pressure creators and users of AI systems to put in place vital and enforceable ethical controls. As discussed in chapter 5, while there is no doubt these programming advancements have improved our lives (search engines, voice recognition, more efficient power systems), there are also ample examples of machine learning gone wrong. With the introduction of autonomous vehicles, AI in weapons systems, and wide-scale adoption of big data algorithms for evaluating the utility of funding and people, our society needs assurances that these systems conform to the best interests of communities.

The quickest route to this has already begun: the development of ethical standards for designing, coding, and implementing AI systems. This is a far cry from Asimov's famous three laws of robotics created in the context of science fiction. We need to be able to trust our phones, cars, and computers today. We need something along the lines of "Everyday Ethics for Artificial Intelligence" being developed by IBM.[12] IBM researchers seek to generate a conversation on a broadly applicable approach to ensuring that AI systems support the aspirations of society. They define five areas of ethical focus:

1. Accountability—a recognition that people create and code these systems and therefore must try to understand the impact these systems will have on the world
2. Value Alignment—being purposeful to build community values and norms into the functions of systems (like ensuring diversity)
3. Explainability—systems should be able to document why they made a given decision or suggested a proposed outcome, in ways that humans understand
4. Fairness—The IBM researchers phrase this as a reduction in bias, which I have argued is impossible, although we would both agree that unintended bias should be minimized and that systems should attempt to be representative of the populations they seek to serve or effect
5. User Data Rights—I'll just go ahead and quote the IBM researchers: "It is your team's responsibility to keep users empowered with control over their interactions."[13]

So, yeah, be humanists.

These standards, which by the way are supplemented with examples and recommended actions to take, would go a long way toward transforming AI into IA—that is, artificial intelligence into intelligence augmentation. IA is actually an old term used around the introduction of cybernetics in the 1950s and 1960s. It was the exploration of how machine capability (later focused on computers) could help people do work better. Today we might just call it increased productivity. However, it is a term that focuses on the desired impact on human capability, not human replacement.

FROM BOOK PALACES TO PALACES OF THE PEOPLE

In chapter 10, I mentioned, albeit briefly, the work of Eric Klinenberg, which supports the importance of public institutions like libraries as vital social infrastructure. This idea that we need libraries and museums to support people in a community, particularly in a time of increasing digitization of the knowledge infrastructure, was backed up by a recent research project in Norway, Sweden, Denmark, Germany, and Hungary.[14] The work of researchers in these countries found that as more and more government services first became digital, and then digital only, there was a greater need for physical and social public spaces like libraries.

In Sweden, Norway, and Finland, national legislation actually mandates libraries to be places where community and societal conversations happen. This included specific provisions to support democratic debate and participation. In a time of increasing digitization, these researchers found an increased need for physical access to the knowledge infrastructure for teaching, learning, creating, and convening as a community. This need for places for communities to gather, to share, and to build trust is essential as we move forward, and an excellent place to begin my discussion of sources in the knowledge infrastructure.

NOTES

1. Kate O'Flaherty, "Beware Zoom Users: Here's How People Can 'Zoom-Bomb' Your Chat," *Forbes*, March 27, 2020, https://www.forbes.com/sites/kateoflahertyuk/2020/03/27/beware-zoom-users-heres-how-people-can-zoom-bomb-your-chat/#c24a2e5618e2.

2. Elizabeth Redden, "'Zoombombing' Attacks Disrupt Classes," *Inside Higher Ed*, March 26, 2020, https://www.insidehighered.com/news/2020/03/26/zoombombers-disrupt-online-classes-racist-pornographic-content.

3. Emma Pettit, "A Side Effect of Remote Teaching During Covid-19? Videos That Can Be Weaponized," *Chronicle of Higher Education*, March 24, 2020, https://www.chronicle.com/article/A-Side-Effect-of-Remote/248319.

4. Oliver Holmes, "Israel to Track Mobile Phones of Suspected Coronavirus Cases," *Guardian*, March 17, 2020, https://www.theguardian.com/world/2020/mar/17/israel-to-track-mobile-phones-of-suspected-coronavirus-cases.

5. Kim Lyons, "Governments Around the World Are Increasingly Using Location Data to Manage the Coronavirus," The Verge, March 23, 2020, https://www.theverge.com/2020/3/23/21190700/eu-mobile-carriers-customer-data-coronavirus-south-korea-taiwan-privacy.

6. Ibid.

7. Byron Tau, "Government Tracking How People Move Around in Coronavirus Pandemic," *Wall Street Journal*, March 28, 2020, https://www.wsj.com/articles/government-tracking-how-people-move-around-in-coronavirus-pandemic-11585393202.

8. Alvaro Puig, "Equifax Data Breach Settlement: What You Should Know," Federal Trade Commission, July 22, 2019, https://www.consumer.ftc.gov/blog/2019/07/equifax-data-breach-settlement-what-you-should-know.

9. "What's Shibboleth?" Shibboleth Consortium, accessed July 31, 2020, https://www.shibboleth.net/index.

10. Phillip Windley, "How Blockchain Makes Self-Sovereign Identities Possible," Computerworld, January 10, 2018, https://www.computerworld.com/article/3244128/how-blockchain-makes-self-sovereign-identities-possible.html.

11. There is a mountain of information on Blockchain and Bitcoin out there, mostly around how to make billions from these, but this is a good document on the basics and how it applies to identity management: "Blockchain Identity Management: The Definitive Guide" (2020 Update), Tykn, June 22, 2020, https://tykn.tech/identity-management-blockchain.

12. "Everyday Ethics for Artificial Intelligence," IBM. accessed July 31, 2020, https://www.ibm.com/watson/assets/duo/pdf/everydayethics.pdf.

13. Ibid.

14. Ragnar Audunson, Hans-Christoph Hobohm, and Máté Tóth, "ALM in the Public Sphere: How Do Archivists, Librarians and Museum Professionals Conceive the Respective Roles of Their Institutions in the Public Sphere?" *Information Research* 24, no. 4 (2019), http://informationr.net/ir/24-4/colis/colis1917.html.

15

Sources

From Swans in Venice to Floods in New Orleans

There are no swans swimming in the canals of Venice, Italy. If you hang on Twitter, you may disagree. You may have been one of the hundreds of thousands of folks who saw @ikaveri's tweet of March 16, 2020: "Here's an unexpected side effect of the pandemic—the water's [*sic*] flowing through the canals of Venice is clear for the first time in forever. The fish are visible, the swans returned."[1] She also included photos of how the canals of Venice had become clear during the Italian pandemic lockdown that had eliminated the throngs of tourists and boats churning up the sediment. One of the photos was of swans in a canal.

Except the swans had not returned. In fact, there haven't ever been swans in the canals of Venice. The picture was of swans in Burano, Italy (about a 40-minute ferry ride away), where they have made their homes in the canals. @ikaveri, whose real name is Kaveri Ganapathy Ahuja, lives in New Delhi, India. She saw the photos of swans on social media, combined them with some photos that others had posted of a quarantined Venice, and tweeted it out. Yet after it was revealed that the story was false, folks got angry. Not at Ahuja, but at the *National Geographic* website for ruining a really good story.

As human beings, we are suckers for good stories. In fact, there is evidence that it is hardwired into our brains. Yuval Harari posits that the ability of humans to create and imagine narratives was a fundamental revolution in our evolution and sits at the core of our society. The problem in a crisis is that narratives are often in flux, can lead us in the wrong direction, and sometimes substitute for facts or be outright wrong.

FROM UNDER CONTROL TO A VERY, VERY PAINFUL TWO WEEKS

In the first months of the coronavirus pandemic, President Trump proffered a number of conflicting narratives—many of them contradicted by reality. The following highlights of the president's rapidly evolving narrative are drawn from Luke O'Neil's March 18, 2020, piece in the *Guardian*, "How Trump Changed His Tune on Coronavirus Again and Again . . . and Again."[2] But this is only one example of what is quickly becoming a genre unto itself. The infection data is from the U.S. Centers for Disease Control,—the federal agency tasked with tracking and controlling outbreaks:[3]

> January 22: "We have it totally under control." Number of infected in the United States: 1.
>
> January 28 (After tweeting a false article about a vaccine in the works): "The risk of infection for Americans remains low, and all agencies are working aggressively to monitor this continuously evolving situation and to keep the public informed." Number of infected: 5.
>
> January 30: "We pretty much shut it down coming in from China." Number of infected: 5.
>
> February 23: "We had 12, at one point. And now they've gotten very much better. Many of them are fully recovered." Number of infected: 15.
>
> February 24: "The Coronavirus is very much under control in the USA." Number of infected: 15.
>
> February 27 (on immigration and the virus): "The Democrat policy of open borders is a direct threat to the health and well being of all Americans. Now you see it with the coronavirus." Number of infected: 16.
>
> March 8: "We have a perfectly coordinated and fine tuned plan at the White House for our attack on CoronaVirus. We moved VERY early to close borders to certain areas, which was a Godsend. V.P. is doing a great job. The Fake News Media is doing everything possible to make us look bad. Sad!" Number of infected: 423.

March 9: "So last year 37,000 Americans died from the common Flu. It averages between 27,000 and 70,000 per year. Nothing is shut down, life and the economy go on. At this moment there are 546 confirmed cases of CoronaVirus, with 22 deaths. Think about that!" Number of infected: 647.

March 10: "When people need a test, they can get a test. When the professionals need a test, when they need tests for people, they can get the test. It's gone really well. Look, the biggest thing that we did was stopping the inflow of people early on, and that was weeks ahead of schedule, weeks ahead of what other people would have done." (Tests were not available). Number of infected: 937.

March 13: "For decades the @CDCgov looked at, and studied, its testing system, but did nothing about it. It would always be inadequate and slow for a large scale pandemic, but a pandemic would never happen, they hoped. President Obama made changes that only complicated things further." Number of infected: 1,896.

March 15: "We are doing very precise Medical Screenings at our airports. Pardon the interruptions and delays, we are moving as quickly as possible, but it is very important that we be vigilant and careful. We must get it right. Safety first!" (No screenings were conducted.) Number of infected: 3,487.

March 16: "We have a problem that a month ago nobody ever thought about." Number of infected: 4,226.

Jon Greenberg of Politifact then picks up the timeline:[4]

March 24: "We cannot let the cure be worse than the problem itself . . . [I would] love to have the country opened up, and just raring to go by Easter." Number of infected 54,433. Number of deaths: 1,000.

March 29: "The peak, the highest point of death rates—remember this—is likely to hit in two weeks. Nothing would be worse than declaring victory before the victory is won. That would be the greatest loss of all." Number of infected: 140,904. Number of deaths: 4,400.

March 31: "This is going to be a very painful—very, very painful two weeks. When you look and see at night the kind of death that's been caused by this invisible enemy, it's—it's incredible." Number of infected: 186,101. Number of deaths: 5,600.

While I am only focusing on the first three months of the pandemic in the United States, here is a little more context. Infections rose to 1.1 million by the end of April, 1.8 million by the end of May, 2.6 million by the end of June, over 11 million by the end of November. And by the end of November, over 250,000 people had died from the virus.

The novel coronavirus as presented went from a nonexistent threat, to a minor inconvenience, to a major pandemic, to a killer of the economy—sometimes back and forth in a matter of hours. Reporters were commenting daily on a new presidential tone or story. The president's live briefings showcased narratives as they were being composed and decomposed in real time leading a number of news outlets to stop covering them live. The worry was that the president's ad libs and statement of belief as fact created a public health danger because of the need for consistent messaging grounded in epidemiology and medicine.

What became evident was that when national government efforts have questionable credibility, it opens members of the public to all sorts of false narratives, conspiracy theories, and outright scams. What is needed, people said, was credible information from authoritative sources. And therein lies the problem. Credibility is no longer determined by authority.

The global pandemic was not the first crisis in which the fault lines of a weaponized and fractured knowledge infrastructure have been evident. In August 2005, Hurricane Katrina flooded New Orleans with rain and a storm surge that sent a 28-foot wall of water toward a city guarded by 23 feet of levees. Fifty-three flood walls protecting the city were breached, and 80 percent of the city was flooded. An inadequate federal response that some have called criminal negligence—a late mandatory evacuation order, flawed reporting, racially charged policing, and a whole host of disastrous environmental decisions over decades that had eliminated protective wetlands—equaled devastation in the form of 1,200 deaths.

In the aftermath of the flooding, New Orleans residents were scattered across the south as they sought shelter. The Federal Emergency Management Agency (FEMA) scrambled to provide temporary housing and services: first in the Superdome, then in the city's convention center after it had descended into squalid and outright dangerous conditions, and finally through trailers and temporary housing. The population of the city in the year 2000 was approximately 485,000. By 2010 it was 348,000.

In the weeks and months following the disaster, citizens grew increasingly frustrated and understandably angry at a lack of consistent and authoritative information coming from the local and state governments. Many New Orleans residents turned to chat rooms and community-run websites to resolve contradictory information coming from traditional sources, including the federal government and the media. Local news websites, such as NOLA.com, allowed communities to come together and share information online directly. Residents were able to hear from multiple sources, including eyewitnesses and other residents, to get a more accurate, complete, and credible picture of the situation in the neighborhoods organized as wards. People who provided the most consistently accurate information became trusted authorities regardless of their titles or affiliations, while traditional authorities lost their credibility. This is but one example of how digital media have turned credibility on its head.

Over the past century, as we have increasingly seen traditional authoritative sources of information become less trusted and more about controlling messages than assuring accuracy, people have become more skeptical—even contemptuous—of those sources. In the extremes of this shift, some turn to ideology for reaffirmation of belief. However, on the large scale, people have turned to sources that they find consistent and open to participation, regardless of source title or position.

Think of how encyclopedias have transformed from authoritative information based on the publisher (*Encyclopedia Britannica*, the *World Book*) to reliable because they are open in terms of who is supplying the information and the changes they make (*Wikipedia*). Think about how we have gone from relying on the nightly news for weather information, to watching online radar to track approaching storms.

This shift in credibility explains much of the polling data and trust we discussed in chapter 10. Who is most trusted? Medical personnel, professors, and librarians whose work can be directly questioned and tested (do I feel better than yesterday? Can I check that against what I find on my own? Can I directly interact and converse with this source?), whereas we have to take the media, politicians, and car salesmen on their own authority—the information too complex or remote for us to interact with.

The current pandemic, as well as the events around Hurricane Katrina, also highlight a paradox that lies at the core of today's increasingly digital knowledge infrastructure: the paradox of self-service in which we seek out our own credible view of a situation and are increasingly, though often invisibly, more reliant upon the sources of that information.

To test the information we receive in a digital world, we now have access to huge array of digital sources. When we want to check the accuracy of the weather report, we can check the radar. When we want to check the president's assertions on the state of the pandemic, we can look at the Johns Hopkins Coronavirus Resource Center,[5] which tracks infections, deaths, and recoveries. By checking these sources, we feel a greater sense of control and can participate in assigning credibility.

The paradox lies in that, in checking multiple sources online, we are actually more dependent on these sources than before. Who says that Johns Hopkins has the right data? In a digital world, who can ensure that the site you pulled up is actually Johns Hopkins and not a scam site made to look like the university? It goes back to the swans in the canals of Venice. We see the picture with our own eyes. We see the picture retweeted and liked by hundreds of thousands of people, and assume that makes a reliability case. But ultimately, we have just replaced one form of unquestioned authority—the news media as an authoritative source—with another—"likes" and retweets. In a situation where getting the right information literally may be the difference between life and death, we need a better way than either trusting a president who pushes unproven medical treatments, or waiting to get sick.

FROM CONSUMER TO PARTICIPANT

There is a presumption built into President Trump's attempt to control the pandemic narrative, and into major parts of the war-shaped information infrastructure itself—that there is a divide between consumers of media and producers of media. It is a divide not over the seemingly special insight an authoritative source may have, but rather over a power dynamic. The source of much of the media available in the knowledge infrastructure has special power to control what is presented.

Our current consumer culture overwhelming seeks to divide people into consumers and producers. In some scenarios, this makes sense: we don't all build our own cars or refrigerators. But in the knowledge domain the distinction is not only muddled; it is deliberately about suppressing voice and power. As I will talk about in chapter 16 on policy, we certainly can look to reform the intellectual property regime that discourages remixing and experimentation. But for this discussion, let me take on the existing producer role we have all been put in by the data-monetization scheme powering much of the digital domain: data provider.

Throughout this book I've provided examples of Americans as data producers. If you have a cell phone, you are providing resellable data to the phone carrier. If you use Facebook or even just use an app that allows for a Facebook login, you are sharing data with Facebook. Smart TV or refrigerator? You are paying part of the purchase price in your viewing and eating habits. I've hammered this one home throughout these chapters, but what should we do about it? Particularly when the data you are providing is being used to make policy decisions? Sticking with the pandemic case for a moment, on April 3rd, 2020, Google announced its COVID-19 Community Mobility Reports web page:

> As global communities respond to COVID-19, we've heard from public health officials that the same type of aggregated, anonymized insights we use in products such as Google Maps could be helpful as they make critical decisions to combat COVID-19.
>
> These Community Mobility Reports aim to provide insights into what has changed in response to policies aimed at combating COVID-19. The reports chart movement trends over time by geography, across different categories of places such as retail and recreation, groceries and pharmacies, parks, transit stations, workplaces, and residential.[6]

In the previous chapter I talked about the fact that these types of services are based on location data that Google, the government, your phone carrier, and advertisers have been collecting from you for well over a decade. But putting this fact in the context that real life-and-death decisions are being made with this data puts special emphasis on the responsibilities of data providers—like yourself.

Over this century in the name of warfare, society has increasingly called upon citizens to sacrifice and to increase personal responsibility. From conscription to rationing to directing industrial war production, contributing to a "greater cause" became a common refrain. As President Roosevelt put it in his April 28, 1942, fireside chat:

> Here at home everyone will have the privilege of making whatever self-denial is necessary, not only to supply our fighting men, but to keep the economic structure of our country fortified and secure during the war and after the war. . . . I know the American farmer, the American workman, and the American businessman. I know that they will gladly embrace this economy and equality of sacrifice, satisfied that it is necessary for the most vital and compelling motive in all their lives—winning through to victory.[7]

It is a common war theme, picked up by President Trump in the outset of the pandemic . . . in at least one of his narratives: "If everyone makes this change or these critical changes and sacrifices now, we will rally together as one nation and we will defeat the virus."[8]

The radio show *On the Media** devoted an entire program to the use of war rhetoric in the breaking pandemic, asking whether it was appropriate to use war metaphors in this case. Brooke Gladstone asked, "Is our reliable stockpile of war metaphors with its explosions and weaponries, soldiers and supply lines, the best vocabulary for framing what's going on?"[9]

The point of this book, however, is that it is really the only metaphor we have . . . worse still, it is not really a metaphor, as all social actions are framed in terms of war and conflict. Politics, campaigns, business competition, and sports are now framed as struggle and, well, warfare. COVID-19 is our "invisible enemy" because after a century of propaganda, military-influenced technology development, and paranoid surveillance public policy that lends itself to monetization through data, all we know how to do is frame and react. The very structure of the way we learn is about fighting. Social cooperation, the need for collective action, and a focus on aspirational statements have devolved into sacrifice for victory, not work for collective action for a better tomorrow.

* Simply one of the best programs on radio. If this sounds like I am sucking up . . . well, a little.

So in this war, and increasingly in wars since Vietnam, what is being called for is increasingly passive consumption and, now, passive production. Our data can be used to save the world—just don't get in the way. Trump demonstrates the impact of his administration by the ratings his briefings receive.

What is needed, however, is a knowledge infrastructure that ensures active and participatory data stewardship. If the data of global citizens is going to be put to use in everything from containing a global pandemic, to supporting community platforms, to setting public policy, then every citizen must be an active steward of their own data.

FROM CENSORSHIP BY NOISE TO CHECKING OUT PRIVACY

There is a concept in media studies called "censorship by noise."[10] It is the reaction of propagandists and political campaigns to a fractured media landscape. When there are few controlled channels of messaging, such as in World War I with the wire services, censorship works by restricting the available messages to the message preferred by those in power. However, in the modern fragmented media landscape, there is a new option. Instead of trying to eliminate an unwanted narrative, drown it out with numerous counternarratives.

The idea of flooding people with apparent choice in order to actually control behavior can have startling effects. The "paradox of choice" is a condition of modern consumers proposed by Barry Schwartz, a psychologist at Swarthmore College.[11] In a well-viewed TED talk he gives the example of buying blue jeans. Many years ago, you only had one or two choices: Lee or Levi's. If the jeans didn't fit, it was the fault of the manufacturer . . . why didn't they make them tighter, or looser, or whatever. But now there are hundreds of jeans manufacturers and styles. Straight fit, slim fit, loose, skinny, bootcut, tapered, slim straight, relaxed. Now if you buy a pair of jeans, the fault if they don't fit is perceived as yours. You think that there are so many options, there must be a perfect fit out there, but you don't have the time to try them all. You failed.

When you have more to choose from, you begin to defer the decision, not wanting to take the time to investigate the possibilities. And when you do make a choice, you realize that (1) There were plenty of opportunities to choose something better, and (2) The person to blame for not having made a better choice is you. Schwartz even goes so far as to suggest that the rise in depression and suicide in the industrialized world is in no small part a result of this paradox.

What does this have to do with data? Well, most likely you suck at maintaining and caretaking the data you share, but is it really your fault? Let's begin with the obvious: you didn't read the EULA. What's a EULA? An End User Licensing Agreement. It's the very long document that you were asked to agree to when you signed up for a Google or Facebook account. These tend to be very long documents written in legalese, primarily about sheltering companies from liability. They say things like you can't have a Facebook account if you are under 13 years old . . . even though *Consumer Reports* in 2011 reported that "there are 7.5 million children under 13 on Facebook."[12] It is also where you signed away your rights to own music you "bought" on Apple's iTunes, as discussed in chapter 2, while also agreeing not to "use these products for any purposes prohibited by United States law, including, without limitation, the development, design, manufacture, or production of nuclear, missile, or chemical or biological weapons."[13]

In order to be a good steward of your data, you must not only read these agreements when signing up, but keep up with updated terms of service, and regularly audit on these services. This is for every social media site and every app you download. How many sites have you signed up for just to "try it," and never went back to close accounts, or see if they have been overrun by spam? Did you know you gave Google and advertisers access to your location data to be used in enforcing quarantine restrictions? Even as companies like Apple have marketed their protection of privacy, and Google and Facebook have implemented plain English annual privacy checkups, how many have not? Also, for good measure, you should probably be reading new rule change proposals from the Federal Communications Commission and the Federal Trade Commission.

The point is obvious. We are living with both the paradox of choice and the paradox of self-service and as our use of data becomes more important, we are more and more reliant on too many parties to even be able to be effective stewards of the data we are sharing. I don't think any of us would object to anonymized data being used to keep people alive, but what else have we signed up for?

What is needed is a trusted data facilitator, and effective regulation of personal data. This is not the first time that the need for such clarity resulted in at least attempted change. After the housing and banking crisis of 2008, it became clear that a lot of people had agreed to loans and credit deals they did not fully understand. This was both at the consumer level (read: you and me) and, clearly, at the level of investors and banking institutions.

The Consumer Financial Protection Bureau was created as part of the 2010 Dodd-Frank Wall Street and Consumer Protection Act and began operations in 2011:

> The purpose of the CFPB is to promote fairness and transparency for mortgages, credit cards, and other consumer financial products and services. The CFPB will set and enforce clear, consistent rules that allow banks and other consumer financial services providers to compete on a level playing field and that let consumers see clearly the costs and features of products and services.[14]

To be sure, the agency has not been without its critics. Under the Trump administration there have been many attempts to weaken the work of the CFPB. However, the idea remains relevant. Just as it was important to create clarity and better control in financial transactions, so too is it essential to safeguard people's use of their own data.

A more distributed solution, however, might be to empower existing local knowledge organizations to do this work—perhaps a trusted local agency staffed with experts in information and with a strong professional ethos and history in privacy and service. Instead of creating more federal structures, we as a nation should empower our state libraries to be public agents charged and resourced to protect citizen privacy—connecting with local public and school libraries to provide direct support and education to local citizens around the topic of data protection. This could be a set of local agencies already in heavy use by key demographics and staffed with knowledge professionals, most with graduate degrees.

Regulation is necessary when there is a fundamental imbalance in a market. When the self-interest of agents in a complex system or market cannot find balance, and when the potential negative impacts on a community are too great, there is an appropriate role for government oversight. It is true of our transportation infrastructure defining the width of road lanes and setting speed limits. It is true of the electrical infrastructure standardizing the shape of plugs, the voltage of current provided, and ensuring that the cost of the service from a given utility is in the best interest of the community. It is true of the banking infrastructure to ensure liquidity and people's deposits. And just as the banking sector was out of balance in 2008, our knowledge infrastructure is out of balance now, with the market's potential profit from the raw resource of personal information outweighing protection of privacy and personal control. The noise of EULAs and apps, and data-for-service, can no longer be tolerated. We cannot say that it is accidental, and we certainly cannot say that it is optimal.

NOTES

1. Natasha Daly, "Fake Animal News Abounds on Social Media as Coronavirus Upends Life," *National Geographic*, March 20, 2020, https://www.nationalgeographic.com/animals/2020/03/coronavirus-pandemic-fake-animal-viral-social-media-posts.

2. Luke O'Neil, "How Trump Changed His Tune on Coronavirus Again and Again . . . and Again," *Guardian*, March 18, 2020, https://www.theguardian.com/world/2020/mar/18/coronavirus-donald-trump-timeline.

3. "Coronavirus Disease 2019 (COVID-19): Cases in the U.S." Centers for Disease Control and Prevention, accessed August 4, 2020, https://www.cdc.gov/coronavirus/2019-ncov/cases-updates/cases-in-us.html.

4. Jon Greenberg, "Timeline: How Donald Trump Responded to the Coronavirus Pandemic," PolitiFact, March 20, 2020, https://www.politifact.com/article/2020/mar/20/how-donald-trump-responded-coronavirus-pandemic.

5. Center for Systems Science and Engineering, "COVID-19 Dashboard," Johns Hopkins University Coronavirus Resource Center, accessed August 2, 2020, https://coronavirus.jhu.edu/map.html.

6. "See How Your Community Is Moving Around Differently Due to COVID-19," Google COVID-19 Community Mobility Reports, last modified July 29, 2020, https://www.google.com/covid19/mobility.

7. Franklin D. Roosevelt, "Fireside Chat 21: On Sacrifice," April 28, 1942, The White House, Washington, D.C., MPEG-3, 32:26, https://millercenter.org/the-presidency/presidential-speeches/april-28-1942-fireside-chat-21-sacrifice.

8. Eli Stokols, Noah Bierman, and Chris Megerian, "Trumps Urges Americans to Avoid Groups, Travel and Restaurants as Coronavirus Crisis Worsens," *Los Angeles Times*, March 16, 2020, https://www.latimes.com/politics/story/2020-03-16/trump-urges-americans-to-avoid-groups-travel-and-restaurants-as-coronavirus-crisis-worsens.

9. Jeet Heer, Nicholas Mulder, Eula Biss, and Bob Garfield, "War, What Is It Good For?" April 3, 2020, *On the Media*, produced by WNYC Studios, podcast, MP3 audio, 51:11, https://www.wnycstudios.org/podcasts/otm/episodes/on-the-media-war-what-is-it-good-for.

10. Peter Pomerantsev, "To Unreality—and Beyond," *Journal of Design and Science*, 6 (October 2019), https://doi.org/10.21428/7808da6b.274f05e6.

11. Barry Schwartz, *The Paradox of Choice: Why More Is Less* (New York: Ecco, 2016).

12. Marc Perton, "Facebook's Zuckerberg Wants to Let Kids Under 13 Onto Site," *Consumer Reports*, May 20, 2011, https://www.consumerreports.org/cro/news/2011/05/facebook-s-zuckerberg-wants-to-let-kids-under-13-onto-site/index.htm.

13. "Apple Media Services Terms and Conditions," Apple, Inc., last modified May 13, 2019, https://www.apple.com/legal/internet-services/itunes/us/terms.html.

14. "Consumer Financial Protection Bureau," Federal Register, accessed August 3, 2020, https://www.federalregister.gov/agencies/consumer-financial-protection-bureau.

16

Policy

From a Commodity to a Public Good

The University of South Carolina's flagship campus has 34,795 students. On March 8, 2020, the students all left for spring break. On March 11, students were given an extra week of break and informed that classes would be offered online until April 5 due to the coronavirus pandemic. They were also asked to not come return to campus until April 5, though essential campus services would remain available to those still in residence halls. On March 15 the university, in accordance with direction from the state's governor, closed the campus to all but essential staff, and students were told not to come to campus at all until at least April 5. On March 19, students were told that the remainder of the semester would be online and that graduation ceremonies were postponed.[1]

What the University of South Carolina, my employer, did was make a seemingly safe assumption—that their students and faculty could connect to the internet. Seemingly safe, in that the Pew Research Center in June 2019 estimated that 9 in 10 Americans use the internet, and roughly three-quarters have access to broadband internet at home.[2] As early as 2013 the White House was reporting that 98 percent of Americans had internet access,[3] so sending students home and assuming they could still connect online, aside from being the only real option, seemed appropriate.

What all of those numbers do assume is that there is no raging pandemic occurring and people are not ordered to stay at home. Those without home internet could no longer get access through their workplaces, or schools, or public libraries. Those that had access to the internet through their smartphones, a growing percentage,[4] also had data caps. All those professors Zooming? It was blowing through the 5-gigs-per-month data plans. While some companies had removed caps by the end of March 2020, others had not.

Some South Carolina students were returning to rural homes with no access, or very expensive services. With 99 percent of all USC freshmen receiving financial aid of some sort,[5] affordability is an issue. And unaffordable internet access isn't just restricted to rural students. For students returning to their homes in New York City:

> A recent report by New York City Comptroller Stringer highlights the problem's severity. Distressingly, 29 percent of households in the New York City—or 2.2 million New Yorkers—are without a broadband connection. In communities of color and disadvantaged populations, this disparity is much more profound. Nearly half of New Yorkers older than 65 live without home internet connections, as do a similar share of all residents in the city's more impoverished neighborhoods such as Tremont in the Bronx and Jamaica, Queens. Broadband, it turns out, isn't so broadly available.[6]

My university was hardly alone in making assumptions about internet connectivity. K–12 schools across the country made the same assumption. Employers made the same assumption. The U.S. government made the same assumption in how it conducted the 2020 Census, sending people codes and URLs in the physical mail instead of paper forms.*

What this crisis has shown is that the internet *is* a utility—a service required for the operation of nearly every other service. What it has also shown, is that the internet *is not* a utility.

Yes, I am very aware that I just contradicted myself. However, the contradiction is not mine; it belongs solely to the Federal Communications Commission (FCC). And yes, this means a quick trip back in time to 2005.

The internet was reaching a crucial tipping point in 2005. The number of users and services on the net had grown to a point where essential services were becoming available *only* on the net. Employers were requiring

* To be clear, people can request paper forms or respond by phone, and field agents will go to physical homes, but the internet is the presumed primary method of response.

applicants to apply for jobs online. Colleges were requiring online applications. Governments at all levels were providing key transparency and policy information only online. The largest growing businesses of the economy were either tech companies or other companies growing through online commerce or services.

The FCC, headed by Republican appointees of the George W. Bush administration, sought to encode so-called net neutrality into policy. Net neutrality is a concept stating that all traffic over the internet should have the same level of priority. Google can't buy a "fast lane" for their service in order to gain a competitive advantage over another search provider like, say, Bing.* If this seems at all familiar, it is a concept that, while not in name but certainly in intention, goes back to the International Radiograph Convention in 1906 that I talked about in chapter 1. Recall that this convention forced telegraph operators to relay messages forward regardless of origin—history matters.

The 2005 FCC policy:

> prohibited internet service providers from blocking legal content or preventing customers from connecting the devices of their choosing to their internet connections. Under this policy, the FCC ordered Comcast in 2008 to stop slowing connections that used the peer-to-peer file-sharing software BitTorrent, which was often used for digital piracy but also had legitimate uses.[7]

However,

> Comcast sued the FCC, arguing the agency had overstepped its bounds. A federal court agreed, ruling that the FCC had failed to make the legal case that it had the authority to enforce the 2005 policy statement.[8]

This led the FCC in 2010, now under a Democratic appointee of the Obama administration, to issue an amended net neutrality order meant to overcome the legal issues and have the same effect. But Comcast sued again, and the same court found fault with the revised order.

In 2014 the FCC proposed to reclassify internet service as a utility, just like the phone lines. While you pay a phone bill, the service is still considered a utility. That means the phone company has to do things like provide a basic

* Obvious Bing joke here.

service to every household regardless of the ability to pay (normally limited to calling 911). It means that the government can force phone companies to install service in rural and urban areas the company may not deem profitable. In essence, just like water service and power service, the government has a say in how a business is operated that overrides (though considers) the drive for profit. It would mean that when a global pandemic consumes the world and the economy, your government and your university and your business could, in fact, assume everyone had an available internet connection.

Except that in 2017 the FCC, now headed by Trump appointee Ajit Pai, undid the utility classification of the internet. The thing that is currently connecting you (maybe) to your school, your family, your food, your work, your government, your news? That's an optional commodity like Netflix and OK Cupid, whose primary regulation is not as a medium of communication under the FCC, but as a business under the Federal Trade Commission.

FROM ACCESS TO MOTIVATION

The division between what one group of people can do with the internet and another has become known as the *digital divide*. It was identified as early as a quarter-century ago when in 1995 the U.S. Department of Commerce published "Falling Through the Net: A Survey of the 'Have Nots' in Rural and Urban America."[9] As early as 1996 the U.S. government attempted to bridge the divide in section 254 of the Telecommunications Act, which established the E-Rate. This legal provision used a portion of the telephone companies' profits to pay for internet connectivity at K–12 schools and public libraries.

What was quickly discovered was that there wasn't just one divide in terms of access to the internet—and therefore, a growing percentage of the knowledge infrastructure—but four. Take an eighth grade teacher in 1997. He or she might get a new internet-connected computer in the classroom. The first type of divide, access, had been solved. However, if the teacher didn't know how to use a computer, or what the internet was (in 1997—pre-Google, pre-YouTube, pre-Facebook, pre-Wikipedia), they were faced with a second digital divide: training. Let's say the teacher did receive training, and now that teacher was asked to use the computer (often for the first time) in front of a group of eighth graders—some of whom may already have greater technical skills than the teacher. This created a third divide: environment. Another example of this divide was shown in a study I supported in the mid-1990s. It showed that in a typical workday, by the time a teacher finished teaching, had lunch, went to the bathroom, and talked with parents, they had 7 minutes left to themselves to search the internet for resources. Never mind that they hadn't started grading or doing lesson plans for upcoming classes.

Then the teacher, or more likely the school, ran into the fourth divide: motivation. The teacher might have access, might have training, might have a supportive environment, and simply not want to use the internet. A quarter-century later we still find plenty of university faculty who have this opinion of the internet in class. Just do a search on "shutting off internet access in the classroom."

If we make the internet truly a utility, and if we continue to push more and more essential services online, we need a policy regime to bridge these four divides: access, training, environment, and motivation. This policy regime has already been laid out, though indirectly, in the preceding chapters. It includes the importance of policies around privacy and data stewardship (as more people go online, they will become part of the data-driven business model). It includes making the internet a utility. It includes supporting the dual roles of libraries in education and the need for physical spaces to foster community cohesion.

FROM OWNED BY DEFAULT TO A CREATIVE COMMONS

As the internet becomes a true utility, it will also require changes in other policies and laws. Changes are increasingly needed around intellectual property and copyright in particular. If we are all producers now, we are also copyright-making machines. This requires changing how we approach the data we produce with tools such as Facebook and Google Maps. We must also look to change policy in regard to the "remix culture." The generation of memes, the sharing of parody songs, and the deluge of phone-created videos to YouTube, TikTok, and Instagram are bringing concepts like copyright infringement and takedown notices ever-closer to our everyday lives.

In 2010, copyright scholars and leading copyright lawyers came together as part of the Berkeley Law School's Copyright Principles Project. The goal was to get a diverse set of interested parties to think about a model copyright law and regulation regime. A remarkably accessible report from the project begins this way:

> A well-functioning copyright law carefully balances the interests of the public in access to expressive works and the sound advancement of knowledge and technology, on the one hand, with the interests of copyright owners in being compensated for uses of their works and deterring infringers from making market-harmful appropriations of their works, on the other. Copyright law should enable the formation of well-functioning markets for creative and informative works that yield benefits for all stakeholders.[10]

The report then laid out 25 recommendations to improve copyright in the United States, most of them about providing greater clarity and more commonsense exemptions to what can be copyrighted, such as:

> Recommendation #16: More elements in copyrighted works than just ideas and information should be excluded from the scope of copyright's protection for original works of authorship.[11]

One of the types of work that would be exempted may surprise you: laws. Aren't the laws of the land already free of copyright? You would think. However, Carl Malamud found out differently.

Malamud, an open-records activist, bought a copy of the Official Code of Georgia Annotated. This is a publication containing not only the laws the state of Georgia passed, but commentary on those laws and links to relevant court proceedings. Malamud copied it onto a website and was promptly sued by the state for copyright violation. The state argued that the actual text of the laws was free and not covered under copyright (never mind that in order to access them, you had to go to a commercial site and agree to never copy or reproduce them). However, the annotated version, the one used by lawyers because they include necessary information like court cases and opinions—even though that too was produced by the state—could be copyrighted,[12] or so Georgia argued. The state claimed that vital information for the enforcement of law and the power of the state, developed through public funding, could be restricted from public dissemination by copyright.

Malamud lost the lawsuit and was ordered to take down the website. But he won on appeal.[13] The State of Georgia then appealed to the U.S. Supreme Court.[14] Malamud won the case in a 5–4 decision.

However, think about what it took to make the legal code of Georgia freely available to all. It took seven years and the intervention of the Supreme Court to decide who owns what. And this, mind you, was without the involvement of multi-billion-dollar corporations. What hope do any of us have?

No wonder the Berkeley project recommended clarifications to fair use (recommendation #4), the creation of a small claims court for minor infractions (recommendation #5), and, perhaps the most relevant here, "Recommendation #15: Copyright law should make it easy for copyright owners to dedicate their work to the public domain."[15] As I covered in chapter 9, all expressions are considered copyrighted at the point of creation. That doodle you just made? The diary entry you just wrote? The finger painting hanging on the fridge? Copyrighted. Unless you did that doodle on the pages of this book—that might be a derivative work. Did you write the diary entry on a work machine? It may be owned by your employer under "work for hire" rules. Was the finger paint a copy of a famous painting? That's just straight appropriation.

The point is, you probably never intended those expressions to be protected under the full power of the copyright law. There are a lot of things that people want widely distributed without concern for payment or protection. Here's the thing: while the copyright law is great at putting things in the system, it makes it rather difficult to get things out. I mean, you could wait until 90 years after your death. But what if you want to put up images, or music, or writings to be copied and used today?

This was the question legal scholar and Harvard Law Professor Lawrence Lessig asked himself. He wanted a practical, legal, and immediate way to circumvent the restrictions of copyright for those who wanted it. What if you wanted anyone to be able to use an image of yours, but only for noncommercial purposes? What if anyone could reuse the file, as long as you receive credit for it? Lessig created a system (and not-for-profit organization) called Creative Commons. It was an intellectual property system built for the internet. You could go to the Creative Commons website, answer a few questions (commercial use or no? Attribution or no? Make changes or no?) and it would spit out a license agreement in plain text and in computer-readable code that you could attach to the item. If you would like to see this in action, do an image search in Google, click on the "Tools" link, and narrow your search by "Usage Rights."

Creative Commons is a great effort toward copyright reform, but we need to go further. If we don't provide greater flexibility and clarity in copyright, we cede yet more power to a smaller number of bigger companies. Now is the time when creativity and greater access to more diverse narratives is needed more than ever. Change happens too fast to be locked in court battles with Disney and Comcast.

FROM CREATIVE COMMONS TO OPEN ACCESS

Reform is great for things that are owned by an individual, but what about things that are already in the public domain, or at least should be? In chapter 12 I mentioned that the National Institutes of Health, a federal agency, spends $5 billion a year on cancer research. In existing copyright law all federally produced intellectual property, including research publications, are automatically in the public domain (with very few restrictions). This means that if government employees were paid out of that $5 billion, the reports they wrote would be open to the world. However, if a researcher at a university or any other place working under a grant or contract writes up and publishes their research, that work can be—and often is—put under copyright. The logic is that the government funding pays for the research and results, but often not the scientific journal papers published based on those results. I use the word logic loosely here, as the results of research that aren't published are pretty close to useless.

This system has been in place for at least the century covered in this book and is made more important by the increased amount of federal dollars invested in research. The system, however, is in crisis. Scholars write papers. They sign away all rights to that paper in return for the prestige of publishing it in a journal, thus helping them with tenure and visibility among peers. The journals then turn around and sell that paper, mostly as part of a journal, back to the university. The cost of these journal subscriptions and the collection of these journals in large online databases is increasing steadily to the point that university libraries can no longer afford to pay for them. And, a large number of these papers are based on funding by the public—a public that has to pay twice; once for the research and again for access.*

While it is easy to blame publishers for this situation, there is plenty of blame to go around. Publishers, after all, have to get the papers reviewed to ensure they are accurate. Publishers have to edit the work, make the articles

* This is the part where I tell you this book was not funded in whole or part through federal dollars.

searchable through descriptive metadata (title, subject, author), and then publish it in a paper journal and/or electronically. And publishers are not making access decisions alone. Many of the journals in question are actually run by editorial boards of scholars and, in some cases, professional associations. One of the largest issues, however, is that academic libraries these days rarely buy individual journal titles, instead buying access to aggregations of thousands of journal titles in so-called big deals. Major library vendors such as Elsevier and EBSCO make money by charging you for nursing journals along with the physics journals, even though your university may not have a nursing school.

This situation is getting better through an Open Access movement. Many journals are now freely available online, their costs greatly reduced by only publishing on the internet. Costs are increasingly covered by research grants directly, or by academics who voluntarily engage in editorial duties like peer review. Over the past three U.S. presidential administrations, there have been increasing pushes and successes in making publicly funded research freely available.[16] As you can imagine, there has also been opposition to this, primarily from scientific publishers.[17]

There are many policies in place that affect the knowledge infrastructure beyond copyright. Patents, for example, have fueled the growth of the technology industry. Many of the initial patents that built the data and media components of the knowledge infrastructure were developed with public funds. There is a whole body of public policy concerning the operation of knowledge industries, from libel law applied to news organizations, to reporting obligations (and prohibitions) in higher education, to privacy surrounding health information. Whole industries have grown around these industries to ensure compliance with these regulations.

At the same time, political and economic philosophies around the role of government and public policy clash, more often than not seeking victory over balance. The weaponized nature of the knowledge infrastructure making social discourse more about winners and losers than about aspirations and the greater good.

It is only by using policy, and technology in an open and shared way, that we can bring together the sources and people of the knowledge infrastructure to continually debate and develop pragmatic solutions—often in the form of compromises. Though we may disagree about the best way to provide health care or save the planet from climate change, we must agree to create an honest and open forum for debate and learning. As a species, it is what we know, how we know it, and sharing that knowledge that has pushed society forward. It is only by investing in the system of learning and knowing that we can now deal with the consequences to the planet and our fellow citizens for that progress.

NOTES

1. Feel free to relive my life in early 2020 here: "Novel Coronavirus (COVID-19)." University of South Carolina, accessed August 3, 2020. https://sc.edu/safety/coronavirus.

2. "Internet/Broadband Fact Sheet," Pew Research Center, June 12, 2019, https://www.pewresearch.org/internet/fact-sheet/internet-broadband.

3. Adi Robertson, "Only 2 Percent of Americans Can't Get Internet Access, but 20 Percent Choose Not To," The Verge, August 26, 2013, https://www.theverge.com/2013/8/26/4660008/pew-study-finds-30-percent-americans-have-no-home-broadband.

4. Monica Anderson, "Mobile Technology and Home Broadband 2019," Pew Research Center, June 13, 2019, https://www.pewresearch.org/internet/2019/06/13/mobile-technology-and-home-broadband-2019.

5. "Office of Undergraduate Admissions: Requirements," University of South Carolina, accessed August 3, 2020, https://www.sc.edu/about/offices_and_divisions/undergraduate_admissions/requirements/for_freshmen/admitted_class_profile/index.php.

6. Emil Skandul, "Why New York City Needs Universal Internet Access," City and State New York, August 7, 2019, https://www.cityandstateny.com/articles/opinion/opinion/why-new-york-city-needs-universal-internet-access.html.

7. "2020 Census Operational Plan," U.S. Census Bureau, last modified February 1, 2019, https://www.census.gov/programs-surveys/decennial-census/2020-census/planning-management/planning-docs/operational-plan.html.

8. Ibid.

9. Eva Johanna Schweitzer, "Digital Divide," *Encyclopedia Britannica*, February 14, 2019, https://www.britannica.com/topic/digital-divide.

10. Pamela Samuelson, "The Copyright Principles Project: Directions for Reform," *Berkeley Technology Law Journal* 25, (2010): 1176, https://www.law.berkeley.edu/php-programs/faculty/facultyPubsPDF.php?facID=346&pubID=221.

11. Ibid.

12. Joe Mullin, "If You Publish Georgia's State Laws, You'll Get Sued for Copyright and Lose," Ars Technica, March 30, 2017, https://arstechnica.com/tech-policy/2017/03/public-records-activist-violated-copyright-by-publishing-georgia-legal-code-online.

13. Joe Mullin, "Appeals Court Tells Georgia: State Code Can't Be Copyrighted," Electronic Frontier Foundation, October 23, 2018, https://www.eff.org/deeplinks/2018/10/appeals-court-tells-georgia-state-code-cant-be-copyrighted.

14. Bill Rankin, "Who Owns the Law in Georgia?" *Atlanta Journal-Constitution*, November 29, 2019, https://www.ajc.com/news/local/high-court-decide-georgia-official-code-free-the-public/ThYNDjOg6V9nPZvtqf59MN.

15. Samuelson, "The Copyright Principles Project: Directions for Reform," 1227.

16. Kelsey Brugger, "White House Formally Invites Public Comment on Open-Access Policies," American Association for the Advancement of Science, February 21, 2020, https://www.sciencemag.org/news/2020/02/white-house-formally-invites-public-comment-open-access-policies.

17. David Kramer, "Scientific Publishers Unite to Oppose Potential Open Access Executive Order," American Institute of Physics, December 20, 2019, https://www.aip.org/fyi/2019/scientific-publishers-unite-oppose-potential-open-access-executive-order.

Society Waypoint

From Chemical Warfare to Healing the World

I have one last trip through time for you. It begins in 1910, four years before the CS *Alert* made her fateful trip. It starts with a German chemist named Fritz Haber and his process for creating ammonia from atmospheric nitrogen and oxygen. While the air you breathe is nearly 78 percent nitrogen and 21 percent oxygen, transforming these from their gaseous form into a liquid and solid was inefficient and costly. The primary means of gathering nitrates—solid salts formed from nitrogen and oxygen—was from bird and bat guano accumulated in caves over thousands of years.* The primary source for these nitrates was South America.

Nitrates are essential in the manufacture of explosives. Gunpower is made from saltpeter, which is a nitrate. Explosives for bombs and warheads are nitrates. After World War I broke out, the naval Battle of the Falkland Islands secured the nitrates of Chile for the British and, later, the Americans. The Germans' primary natural source was cut off. They turned to Fritz Haber.

Haber not only helped the German Empire invent and then industrialize ammonia production—the first step to manufacturing nitrates; he turned his chemical genius to the development of chemical weapons. First chlorine gas, a choking agent that suffocated the enemy, then mustard gas, a sulfur blistering agent that burned the skin, scarred the lungs, and, as it turns out, prevented

* If that sounds extremely odd, do recall that all oil and plastics come from liquified organic compounds (read, the dead bodies of plants and animals) compressed underground.

the production of white blood cells in the body—leaving survivors open to infections like the 1918 Spanish flu.

As doctors studied this unexpected side effect of mustard gas in World War I, and then in World War II when the U.S. Army deliberately exposed African American soldiers to the blistering agent, they discovered the cause of the neutropenia—a low white cell count. Mustard gas was an alkalizing agent. It and other alkalizing agents interrupted DNA replication in rapidly growing cells such as stem cells deep in the marrow of bones that make white blood cells and the platelets your body uses to stop bleeding.

It turns out there is another type of rapidly growing cell that functions on white blood cells: cancer. Lymphomas are cancers of the immune system. In both lymphoma and leukemia—cancers of the blood—there is no single point of disease, as is the case in solid tumor cancers like lung, breast, or brain cancer. In the first half of the twentieth century the primary treatment for cancer was surgery. That is, you cut it out. You couldn't cut out leukemia or lymphoma, so they were particularly deadly. Doctors discovered that controlled exposure to agents like mustard gas, in the form of an intravenous drip, could slow, and even stop, some blood cancers. Chemotherapy went from a theory to a reality.

Over the next decades all sorts of chemical and biological agents were found to kill all sorts of cancers. Chemotherapy took the place of surgery as the frontline treatment of most cancers. Cancers such as Hodgkin's lymphoma became curable diseases. A derivative of mustard gas, mechlorethamine, is still part of several chemotherapy "cocktails." It is also instrumental when chemo fails and patients suffering from cancers like lymphoma, leukemia, and myeloma need bone marrow transplants. In these cases, it is the very marrow of a patient's bones that is producing the cancerous cells. A bone marrow transplant kills off the patient's blood-making organ and replaces it with either a "clean" version of their own marrow, or marrow from a donor.

It was for this reason that on February 23, 2014, I was injected with a lethal dose of mechlorethamine—lethal in that the chemical would kill all of my bone marrow. Unable to make any more red blood cells to carry oxygen throughout my body, I would die of asphyxiation. Unable to make platelets, my blood vessels would leak, and I would bleed out. Unable to make white blood cells, I would succumb to the most minor of infections. The injection would be lethal, except that once the new form of mustard gas had done its job, a new set of stem cells would be injected through a catheter leading to my heart, find their way to the core of my bones, and begin a new blood-making operation.

I am alive today because Fritz Haber developed new ways to kill soldiers on the battlefields of the Western Front some 106 years ago.

It is this path through history that began this book. It was a fascination with how the waging of war shaped and saved my life that began my look into how it shaped data and media and the world around us. It was on this story that I planned to begin this book. I thought I had it written before even the first paragraph of the first page. But it is not the story I thought it was. After writing this book, the meaning of this story has changed for me.

In 1918 at the end of World War I, Fritz Haber received the Nobel Prize in chemistry. He did not receive this work for the development of chemical weapons and the lives they took, but because of his process for fixing nitrogen from the air as ammonia, and then ammonia nitrate. You may well have heard of ammonia nitrate because, while it is still highly explosive in quantity, it is a common plant fertilizer. From massive farming operations to backyard gardens, nitrogen fertilizer makes plants grow faster and with greater yield. It is the invention of artificial nitrogen fertilizers that is credited for the Green Revolution of the 1950s that helped China and India grow more productive crops for an exploding population.

And so now, I ask myself, what if Fritz Haber had not turned his attention from growing plants to killing people in a War to End all Wars? Would I still be alive? Would a different chain of events from the killing fields of France to a hospital room in Syracuse still have saved my life? As a cancer patient I owe a lot to war, from chemotherapy to advances in radiation treatments. Yet did it take war, with all of its costs, to save lives?

We, of course, can never know. There is some evidence that the push to save lives is a strong enough driver to produce chemotherapy and cancer cures. The new wave of cancer treatments based on genetic manipulation of a patient's own immune system, so called CAR-T, was not developed on the battlefield. The rhetoric of war and the exploitation of personal data did not drive that work. Unlike the Black soldiers marched into closed chambers of mustard gas in World War II, or the unwilling Jewish victims of medical experimentation by the Nazis, the clinical trials for CAR-T were conducted on volunteers who gave access to their bodies and their medical histories through a process of informed consent. The doctors and researchers shared their results openly with other scholars through publications and conferences.

Yet, even here we see how patents on processes and gene sequences can slow the roll out of new treatments. Those scholars developing and refining the CAR-T process still relied on massively expensive journal database subscriptions, out of reach of doctors in huge parts of the world.

CAR-T is an expensive solution to saving lives. Treatments can cost millions of dollars. Doctors must genetically modify each patient's immune system individually. The developed cure is only available to a single patient. How will we pay for these treatments? How can insurance companies stay in business? How can the uninsured ever hope to afford their own lives? Our knowledge infrastructure, framed to cover conflict and sensational miracle cures, clouds the picture and makes necessary pragmatic compromise on health care nearly impossible. And yet we desperately need a balanced, diverse, and equitable knowledge infrastructure to heal the earth, support every citizen, and invent the future.

FROM HISTORY TO THE FUTURE

There is my case: That the knowledge infrastructure is important, and that the people, technology, information sources, and policy that make up this structure have been shaped by the twin fields of data and media. That our understanding of data and media has been crafted by a century of war and conflict. That our present means of knowing the world, finding meaning and power within it, is compromised by an information landscape focused on monetizable data and a media landscape that seeks to concentrate the control of national and international narratives in the hands of a few. That the global COVID-19 pandemic, and disasters from hurricanes to terrorist attacks, clearly demonstrate the fractured nature of the knowledge infrastructure and a need to change. Lastly, that reframing our development of the infrastructure requires acknowledging that what we know is socially constructed, and we must adopt diversity in all its forms; return control of data to the people; place ethical constraints on artificial intelligence; reform copyright law to provide greater clarity and creative access; and ultimately reject the deterministic belief that data alone can guide human and social progress.

It is an incomplete case, I know. Economics, politics, sociology, and the humanities all need to add their perspectives, correct my errors, and add to an agenda of change. But at the base, we need a ready system for them to do so. This means media open to complex ideas, universities driven toward truth rather than profit, and technology companies that seek to trigger our better natures instead of our dopamine receptors. We need a government more interested in our well-being than in crafting messages about our well-being.

I was once called a pragmatic utopian. It is a title I carry with honor. As I said at the outset of this book, this is an optimistic piece. I believe as a society we have the power and the will to transform *how* we know into a more diverse and open system. I believe that data, while an essential part of understanding the universe, is best seen as a tool among many to understand the human experience. I believe that what drives us every day as a species is less about profit, and more about meaning. I believe that the narratives that drove the development of both data and media—narratives of connection and learning and seeking common purpose—were not naive, just distorted by a world constantly at war. And I believe that if we retake those narratives, we can avoid those wars altogether.

The end?

Excursus

On the Joys of Writing and How to Check My Facts

The *Oxford English Dictionary* defines excursus as "1. The Latin word is used by editors of the classics to signify: A detailed discussion (usually in the form of an appendix at the end of the book, or of a division of it) of some point which it is desired to treat more fully than can be done in a note. Hence occasionally applied to a similar appendix in other works." And "2. A digression in which some incidental point is discussed at length."[1] For me, it is an opportunity to talk a little about my writing process, a little on scientific methodology, and a lot about the amazing work of others on which I built this book.

It is hard to express the joy I had while writing this book. For me the process itself is fascinating. Every chapter is like a puzzle. How does the narrative progress, where do the examples go, what to leave in, what to leave out? And these puzzles are wrapped in the larger puzzle of how the book itself fits together. All of this in a process I see as akin to sculpting—setting up a first draft "wire frame," and then going back and adding, removing, and shaping and shaping and shaping.

The joy is also in inviting others into the shaping process. It began when I emerged out of a fevered bout of hypergraphia to ask if what I had started mattered. Will it make sense? Will anyone care? I posted these questions in timid conversations with my wife and close friends. Next came the first cautious sharing of draft chapters. And then, the more emboldened book proposal to editors and potential publishers. The joy of working with others then became commenting on drafts, rambling conversations on history, spirited debates on the conclusions, the pushing and pulling on which ideas deserve more attention, and which are simply too weak to stand.

I have nearly lost myself in the writing of others. This work stands on the shoulders of those scholars and authors who went through their own process. I am indebted to the historians on whose primary research I base my work and analysis. I am indebted to the journalists, librarians, information scientists, and the scholarly community that created the corpus I used. And this brings me to the point of this excursus. If you want to check my work, or more importantly, lose yourself in the work of others on these topics, here is my recommended reading list.

This following books, documents, videos and—well—stuff were instrumental in writing this book. It is not a complete citation list, but rather materials that served as information and inspiration and, for you, places to dig deeper into topics I only presented at a surface level. Also note that, when I say helped me in formation, at times that was because I strongly disagreed with them.

I highly recommend the work of James Burke, journalist and historian of science and technology. While I don't cite him once here, his view of history and presentation of the past as themes shaping the present have been transformative in my life since I first encountered his *Day the Universe Changed* BBC series broadcast on public television in the United States in 1986. His television work also includes *Connections.* In terms of books, I have been fascinated and influenced by his attempts to represent his networked conception of history in a written linear format. Of particular note are:

- *The Pinball Effect: How Renaissance Water Gardens Made the Carburetor Possible and Other Journeys Through Knowledge* (Boston: Little, Brown & Company, 1996)
- *Circles—Fifty Round Trips Through History. Technology. Science. Culture* (New York: Simon & Schuster, 2000)
- *The Knowledge Web* (New York: Simon & Schuster 2001)

If you wish to explore the history of data and information science, go read (like right now) Kathy Peiss's *Information Hunters: When Librarians,*

Soldiers, and Spies Banded Together in World War II Europe. It is a very deep dive into the largely unrecognized work of librarians, archivists, and booksellers in World War II. It is told from the perspectives of the people who lived the work. It is also, as a professor of library science, the best bit of propaganda for librarianship I have ever run across.

Other key reads in terms of history would be the work of Yuval Harari. *Sapiens* may be one of my favorite books of all time, but also check out *Homo Deus*. His conceptions of how mankind evolved because of the ability to believe in the imaginary (like corporations and money) was revelatory. If you would like to know more about the propaganda campaigns through history, I highly recommend (and leaned heavily upon) Philip Taylor's 2003 *Munitions of the Mind: A History of Propaganda*. I focused on the sections from World War I to today, but it deserves a read as the rich history of propaganda back to ancient times. Also recommended is Jonathan Winkler's 2013 *Nexus: Strategic Communications and American Security in World War I*. Lastly in terms of history, read Gordon Corera's 2017 *Cyberspies: The Secret History of Surveillance, Hacking, and Digital Espionage*.

Definitely one of my favorite books of all time is John Barry's *The Great Influenza*. It is unfortunate that the COVID-19 pandemic was the occasion to reread it, but I'm so happy I did. His interweaving of the personal with the epic, with the history of war and medicine and disease, is simply jaw-dropping.

I am indebted to Seth Stephens-Davidowitz's 2017 *Everybody Lies: Big Data, New Data, and What the Internet Can Tell Us about Who We Really Are*. Let me simply say it provided a great number of examples on the thinking of some data scientists. As did Cathy O'Neil's *Weapons of Math Destruction: How Big Data Increases Inequality and Threatens Democracy*, though in a very different way. I think O'Neil's book should be required reading by anyone who uses the internet.

The Pew Research Center and Lee Rainie are national treasures.

The radio program *On the Media* is like a constant running national media conscience that should be attended to.

If you are up for a scholarly adventure, read the works of John Holland on complexity theory, Thomas Kuhn on scientific revolutions, and John Dewey on education reform. When you are done reading them, I think you deserve a PhD.

I thank the authors of all of the sources cited here, from books to YouTube videos. You truly demonstrate the reality that the richest learning comes from the richest sources.

NOTE

1. *Oxford English Dictionary Online*, s.v. "Excursus," accessed August 6, 2020, https://www.oed.com/view/Entry/65956?redirectedFrom=excursus&.

Index

About the Author

R. David Lankes is a professor and the director of the University of South Carolina's School of Information Science. Lankes has always been interested in combining theory and practice to create active research projects that make a difference. His work has been funded by organizations such as the MacArthur Foundation, the Institute for Library and Museum Services, the National Aeronautics and Space Administration, the U.S. Department of Education, the U.S. Department of Defense, the National Science Foundation, the U.S. Department of State, and the American Library Association.

CPSIA information can be obtained
at www.ICGtesting.com
Printed in the USA
LVHW092022260421
685609LV00001B/16